PREPARATION FOR
GENERAL CHEMISTRY

PREPARATION FOR
GENERAL CHEMISTRY

HEWITT GLENN WIGHT
Professor of Chemistry
California Polytechnic State University

DAVID G. WILLIAMSON
Associate Professor of Chemistry
California Polytechnic State University

McGRAW-HILL BOOK COMPANY
New York   St. Louis   San Francisco   Düsseldorf   Johannesburg
Kuala Lumpur   London   Mexico   Montreal   New Delhi
Panama   Rio de Janeiro   Singapore   Sydney   Toronto

PREPARATION FOR GENERAL CHEMISTRY

Copyright © 1974 by McGraw-Hill, Inc.
All rights reserved.
Printed in the United States of America.
No part of this publication may be reproduced,
stored in a retrieval system, or transmitted,
in any form or by any means, electronic, mechanical,
photocopying, recording, or otherwise,
without the prior written permission of the publisher.

234567890KPKP7987654

This book was set in Palatino by Textbook Services, Inc.
The editors were Brete C. Harrison, Robert H. Summersgill,
and Shelly Levine Langman;
the designer was Nicholas Krenitsky;
and the production supervisor was Sam Ratkewitch.
The drawings were done by Vantage Art, Inc.
Kingsport Press, Inc., was printer and binder.

Library of Congress Cataloging in Publication Data

Wight, Hewitt Glenn.
    Preparation for general chemistry.

    1. Chemistry. I. Williamson, David G., joint
author. II. Title.
QD33.W68   540   73-13623
ISBN 0-07-070165-2

CONTENTS

Preface			xiii
CHAPTER 1		THE SCIENCE OF CHEMISTRY	2
		Chapter Topics	2
	1-1	Introduction	2
	1-2	More New Terms	4
	1-3	The Scientific Method	6
	1-4	The Metric System	8
	1-5	Conversion of Units	14
	1-6	Temperature Conversion	18

PREPARATION FOR
GENERAL CHEMISTRY

	1-7	Density	22
	1-8	Specific Gravity	25
		Glossary	26
		Self Test	28
		Answers	29
		Exercises	30
CHAPTER 2		THE ELEMENTS	36
		Chapter Topics	36
	2-1	Introduction	36
	2-2	Broad Classifications of the Elements	37
	2-3	Names and Symbols of the Elements	38
	2-4	Some Common Elements	42
	2-5	The Periodic Table	43
		Glossary	47
		Self Test	48
		Answers	49
		Exercises	49
CHAPTER 3		THE ATOM	54
		Chapter Topics	54
	3-1	Brief History	54

	3-2	The Atomic Theory	55
	3-3	The Structure of the Atom	58
	3-4	The Atomic Structure of the First 20 Elements	65
	3-5	Atomic Weight and Isotopes	67
	3-6	Atomic Structure and the Periodic Chart	70
		Glossary	72
		Self Test	74
		Answers	75
		Exercises	76
CHAPTER 4		COMPOUNDS	82
		Chapter Topics	82
	4-1	Common Compounds	82
	4-2	Particles Involved in Compound Formation	85
	4-3	Covalent Bonding	86
	4-4	Ionic Bonding	88
	4-5	Compounds with both Electrovalence and Covalence	91
		Glossary	94
		Self Test	95
		Answers	96
		Exercises	98

PREPARATION FOR
GENERAL CHEMISTRY

CHAPTER	5	NOMENCLATURE AND CHEMICAL FORMULAS	102
		Chapter Topics	102
	5-1	Nomenclature of a Few Simple Compounds	102
	5-2	Writing Formulas of Chemical Compounds	104
	5-3	Variable-valence Metallic Ions	106
	5-4	Familiar Substances	108
		Glossary	109
		Self Test	110
		Answers	110
		Exercises	111
CHAPTER	6	MATHEMATICS AND CHEMISTRY I	116
		Chapter Topics	116
	6-1	Mathematical Calculations and the Study of Chemistry	117
	6-2	Calculation of Molecular Weight	118
	6-3	Calculation of Percentage Composition	120
	6-4	The Mole Concept	123
	6-5	Empirical and Molecular Formulas	130
		Glossary	138
		Self Test	138
		Answers	141
		Exercises	141

CHAPTER 7	CHEMICAL REACTIONS AND CHEMICAL EQUATIONS	148
	Chapter Topics	148
7-1	Chemical Reactions	148
7-2	Writing a Chemical Equation	150
7-3	Types of Chemical Reactions	152
7-4	Examples of Equation Writing	153
	Glossary	161
	Self Test	162
	Answers	164
	Exercises	164
CHAPTER 8	MATHEMATICS AND CHEMISTRY II	170
	Chapter Topics	170
8-1	Calculations Based on Chemical Equations	170
	Glossary	184
	Self Test	185
	Answers	186
	Exercises	186
CHAPTER 9	SOLIDS, LIQUIDS, AND GASES—THE KINETIC-MOLECULAR THEORY	198
	Chapter Topics	198

PREPARATION FOR
GENERAL CHEMISTRY

9-1	Solids	198
9-2	Liquids	201
9-3	Gases	201
9-4	The Kinetic-Molecular Theory	202
9-5	Changes of State	206
	Glossary	207
	Self Test	209
	Answers	210
	Exercises	211
CHAPTER 10	**SOLUTIONS**	**214**
	Chapter Topics	214
10-1	The Concept of a Solution	214
10-2	Concentration of Solutions	215
10-3	Molarity and Other Concentration Units	215
10-4	Percent by Weight	222
	Glossary	226
	Self Test	227
	Answers	228
	Exercises	229
CHAPTER 11	**ACIDS, BASES, AND SALTS**	**236**
	Chapter Topics	236
11-1	Acids	236

11-2	Bases	239
11-3	Salts	240
11-4	Reactions of Acids, Bases, and Salts	241
11-5	Ionic Equations	244
11-6	Normality	246
	Glossary	249
	Self Test	251
	Answers	253
	Exercises	254

CHAPTER 12 EQUILIBRIUM 262

	Chapter Topics	262
12-1	Dynamic Equilibrium	262
12-2	Shifting the Position of Equilibrium	264
12-3	Strengths of Acids and Bases	265
12-4	pH: The Acidity or Basicity of a Solution	269
	Glossary	270
	Self Test	271
	Answers	272
	Exercises	273

CHAPTER 13 CALCULATIONS DEALING WITH GASES 280

	Chapter Topics	280
13-1	Introduction	281

	13-2	The General Gas Law	281
	13-3	Changes in Temperature, Pressure, and Volume	288
	13-4	The Volume of 1 Mole of a Gas (Molar Volume)	296
	13-5	Weight-Volume Problems	296
		Glossary	297
		Self Test	298
		Answers	300
		Exercises	300
APPENDIX I		EXPONENTIAL NOTATION (SCIENTIFIC NOTATION)	307
APPENDIX II		ALGEBRAIC MANIPULATIONS	309
APPENDIX III		CONVERSION FACTORS	311
APPENDIX IV		ELECTRONIC CONFIGURATION OF THE FIRST 36 ELEMENTS	313
APPENDIX V		SHAPES OF ORBITALS AND MOLECULES	316
APPENDIX VI		ADDITIONAL INORGANIC NOMENCLATURE	318
APPENDIX VII		ANSWERS TO SELECTED EXERCISES	320
Index			331

PREFACE

This brief text has been written for those college students who believe they are going to have a difficult time studying general chemistry, that is, students who find themselves faced with the necessity of taking a course for which they are not well prepared. They are generally intelligent people, but they tend to have little interest in, or aptitude for, the study of chemistry.

In writing this text our objectives were to

Write a concise book for a brief course which might be presented in either one quarter or one semester.

Include topics that provide a necessary background for further study.

Include many examples of common occurrences.

Assist in developing students' ability to carry out mathematical manipulations.

Use an elementary approach throughout the book. (The student who progresses to more sophisticated courses will frequently encounter more sophisticated definitions as level of comprehension increases.)

We would like to point out certain features of this text:

Each chapter is introduced by a section that alerts the student to the main concepts to be learned in studying the chapter.

A Glossary is provided at the end of each chapter, which summarizes the definitions of new terms.

At the end of each chapter a Self Test and Answers are provided to aid the student in evaluating his understanding. Many Exercises are also included for practice. (Answers to every third exercise are given in Appendix VII.)

It is hoped these features will be useful aids to students in learning chemistry.

An effort has been made to design Examples and Exercises that utilize commonly encountered situations. Hopefully these will emphasize the possible frequent use of principles that are presented.

The sequence of chapters may be varied if desired; in particular, Chapter 13 might easily follow Chapter 9 without loss of continuity.

Throughout, an effort has been made to emphasize basic concepts without camouflage.

HEWITT GLENN WIGHT
DAVID G. WILLIAMSON

PREPARATION FOR
GENERAL CHEMISTRY

THE SCIENCE OF CHEMISTRY

CHAPTER TOPICS

In reading this chapter, you should learn what is meant by the term *chemistry*

The definitions of some important terms used in chemistry (see the Glossary at the end of the chapter)

The meaning of the *scientific method* of thought

The fundamental units of the *metric system*

How to use *scientific notation*

How to convert one unit of the metric system into another unit of the metric system

How to convert metric system units into *English system* units (and vice versa)

How temperature is expressed, and how to convert *Fahrenheit* temperature into *Celsius* temperature (and vice versa)

What *density* is, and how to do density problems

What *specific gravity* is, and how to do problems involving specific gravity

1-1 INTRODUCTION

Chemistry is a fascinating subject. Many people, unfortunately, have been frightened by wild tales of the difficulty of this subject and therefore have missed one of the greatest intellectual adventures in the world today. As you are reading this book, look around you; whatever you see is composed of chemicals. Some substances, such as living plants and animals, are composed of very complex chemicals; other substances, such as water and salt, are composed of rather simple chemicals. In the course of study that we are beginning, we shall deal mostly with the simpler substances because they are easier to understand. If you go on to advanced work, you will learn more about the complex chemicals. As you master an understanding of this fascinating world in which we live, you will find that it becomes increasingly interesting.

A part of the study of any science involves learning the language used in that science; therefore we shall commence our study of chemistry by learning the meanings of some words that will be used frequently. Although some of these words may be unfamiliar to you at first, if you accept every opportunity to use them correctly, you will soon master the language of chemistry. One of these words is *matter:* Matter may be defined as any real substance; any such substance is attracted by gravity and therefore has *weight*. This means that any real material is some form of matter; if you can touch it, weigh it, or smell it, it is *matter*.

Now that we know what matter is, we can define *chemistry*. Chemistry may be defined as the study of the structure of matter, the changes that matter will undergo, and the associated energy changes. In other words, chemistry deals with an understanding of the nature of all those physical materials that make up the world in which we live. The average person, in everyday living with materials, acquires a considerable amount of chemical knowledge; in this book we shall delve more deeply into the structures of "everyday" materials and the changes that they undergo. By understanding the nature of physical materials, we can more readily *control* the changes that they will undergo. If we can control these changes, we can cause them to be helpful to us.

Thus, this science of chemistry can affect the life of every living person. In this book, we shall deal particularly with the structure of matter and the changes that matter will undergo; in more advanced courses, the associated energy changes are studied in more detail.

1-2 MORE NEW TERMS

Additional words with which you should become familiar before going further are defined briefly here. A *substance* is usually taken to mean a *pure*, or *single kind*, of material; some examples of pure substances are pure water, aluminum, salt, copper, and sugar. In this sense, you would not refer to a combination of salt and sugar as a single substance, but as a *mixture of substances*. The *properties* of a substance are those characteristics by which the substance may be identified. If we were describing a person, we might give the color of that person's eyes, his (or her) height, his weight, his complexion, his dimensions, and so forth. Similarly, to identify a substance, we would give its color, odor, density, boiling point, and any other characteristics that we might be able to measure. If any one property of a pure substance is different from the corresponding property of another pure substance, then these are *two different substances*.

An *element* is a substance that cannot be decomposed into simpler substances by ordinary processes; sulfur, copper, silver, oxygen, mercury, and many other familiar substances are elements. Prior to 1900 it would not have been necessary to add "by ordinary processes" to our definition. Since about 1900, however, men have learned to convert certain elements into other elements. This involves a type of change called a *nuclear change,* which is quite different from any ordinary chemical process.

A *compound* is a substance that is composed of two or more elements chemically combined in a definite proportion by weight. Simple compounds are such things as water, salt, baking soda, carbon dioxide, and sulfuric acid. More complex compounds are such things as sugar, starch, fat, protein, and other compounds that are generally found in living organisms.

A *mixture* is a material that can be separated into its different parts by mechanical processes. This means that the different substances in a mixture are *not chemically combined,* but are *mixed* together. If you were to take table salt and mix it with sand, the result would be a *mixture.* You could separate the salt from the sand by dissolving the salt in water and filtering out the sand; if the water solution were then evaporated, the salt would remain and you would have separated the salt and the sand. In the process, you would not have carried out any *chemical reaction.* In a mixture the proportions of the substances that are mixed can be varied. In a mixture of salt and sand, more salt may be added, but one would still have a salt and sand mixture. In a compound, on the other hand, the proportions are fixed; pure water, for example, is always 88.8% oxygen by weight and 11.2% hydrogen. The composition of pure water is always the same and is independent of the source of the water.

A *homogeneous* material is one that is uniform throughout. Any pure element or compound is a homogeneous substance. Some mixtures, such as solutions, are homogeneous, but such a mixture as salt and sand (just discussed) is not homogeneous. These nonhomogeneous mixtures are said to be *heterogeneous*.

1-3 THE SCIENTIFIC METHOD

The science of chemistry as we know it is relatively new. Before the seventeenth century, people had very different ideas concerning the structure of matter. The ideas that we have today have been developed by the use of the *scientific method*, which may be said to involve four steps:

1. A careful observation of *facts*
2. The statement of *laws*
3. The invention of *theories*
4. The testing of *theories*

To consider the scientific method in more detail, let us examine each step more fully. The first step, a careful observation of *facts*, means that the good scientist is more careful in his observations than most people. Whenever possible, he will make accurate measurements rather than mere casual observations. There are thousands of people working in scientific laboratories every day, who are making very careful observations and measurements in order to learn more about the nature of matter.

After a large body of facts has been gathered dealing with a certain topic, it is often possible to make a generalized statement that summarizes these facts briefly. This generalized statement is called a *law*. When one or more laws dealing with a certain topic have been clearly stated, it is sometimes possible to form an imaginative picture that will explain why these laws are observed. This imaginative picture is originally called an *hypothesis*. An hypothesis may also be said to be an attempt to explain the existence of laws by using a model which involves familiar concepts. If an hypothesis is investigated further and withstands this more detailed investigation, it may be further dignified by calling it a *theory*.

An hypothesis, or a theory, is tested by making predictions as to what might be expected to happen in cases that have not been observed as yet. If a theory is able to correctly predict facts that have not been previously observed, we begin to have confidence in our theory, and we feel that it may be a reasonably accurate picture of what is actually happening.

It should be mentioned, however, that in actual practice the procedures used by scientists often do not follow exactly the steps of the scientific method in the order indicated. A genius may have a burst of inspiration and accurately deduce a correct hypothesis, with only a few facts as a basis for reasoning. The remaining steps in the scientific method might then be filled in by others. Additional variations in the sequence of the steps of scientific method may also occur, but the four steps just discussed are usually involved at one stage or another in genuine progress. (In actual practice we often consult the literature of the science that we are interested in to see what other people have done in the area before we start making measurements ourselves. If we have reason to doubt or disbelieve the results that someone else has published, we always reserve the right to make the measurements ourselves.)

As we study and learn the many facts that have been observed, the carefully stated laws that summarize these observations, the theories that help us to understand and explain these laws, and the testing of these theories, we shall often refer back to the *scientific method*. As we practice thinking in these terms we shall develop a most useful tool for the solution of many types of problems that come into our lives.

We should recognize that we expect scientific *laws* will be the same in any part of the world or universe that we may visit in the future. *Theories*, however, are the products of man's imagination, and as such are subject to change as new facts are learned.

1-4 THE METRIC SYSTEM

In applying the first part of the scientific method, it is necessary to make careful measurements. To do this, chemists and other scientists have found it convenient to use a system of measurements called the *metric system*. This system was set up by a committee of scientists under the direction of the French National Assembly in about 1790. These scientists were dissatisfied with the *systems* of units; they recognized that it would be much easier to make calculations dealing with measurements if all units used were multiples of 10. They also wanted to develop a system with a simple, direct relationship between units of length, volume, and weight.

Let us consider some of the shortcomings of the English system of units in order to help us to recognize why scientists felt that a new system needed to be established. This English system is typical of a system that was never carefully planned but just "grew." Let us first consider the units that are used for measuring length: The *inch* is a satisfactory unit considered by itself, but there are 12 inches in 1 foot. This means that it is necessary to multiply the number of feet by 12 and then add the number of odd inches to find how many inches there are in a longer measurement, such as 4 feet 5 inches. Also, as we know, there are 3 feet in 1 yard and 5,280 feet in 1 mile. Calculations involving these units are cumbersome and time-consuming.

Another shortcoming of the English system is the lack of a simple relationship between units of length and units of volume. For example, 1 United States liquid quart is 0.033421 cubic foot, which is an awkward figure to use in calculations. We could cite many other examples of the shortcomings of the English system of units, but the above examples should help us see that an entirely different system might greatly simplify calculations. (Having observed some of the awkward features of the English system of measurements, it is easy to understand why legislation passed by the United States Congress recommends that the United States change over to the metric system.)

In setting up the metric system, a fundamental unit of *length* was established; this unit of length is called a *meter*. The meter was originally intended to be one ten-millionth of the distance from the equator of the earth to one of the poles, or one ten-millionth of the earth's quadrant. In order to have a standard that might be referred to readily, it was decided to create one by making two very careful scratches on a metal bar; the distance between these two scratches was made to be as close to one ten-millionth of the earth's quadrant as was possible at that time (1790). The distance between these two scratches was then adopted as the standard unit of length, the *meter*. The metal bar with the two precious scratches, now called the *international prototype meter*, is carefully preserved in the International Bureau of Weights and Measures near Sèvres, France. In recent years it was necessary to make a more precise fundamental definition of the length of a meter for very accurate work. This definition was made using spectroscopic methods.[1]

A careful comparison has been made between the English system of units and the length of the meter. It has been found that the meter (the distance between the two scratches on the metal bar) is equal to *39.37 inches.*

Other metric system units of length were also established: One-tenth ($\frac{1}{10}$) of the meter was called one *deci*meter. One one-hundredth ($\frac{1}{100}$) of the meter was called one *centi*meter. One one-thousandth ($\frac{1}{1000}$) of the meter was called one *milli*meter. Additional units of length were also established, as shown in Table 1-1 (together with their abbreviations). Note the prefixes; these are also used in units of volume and weight.

In dealing with the metric system, it is frequently convenient to use the method of writing numbers as *powers of 10,* sometimes called *scientific notation.* This involves first a knowledge of writing numbers as *powers,* as illustrated in Table 1-1. Thus

$$10^2 = 10 \times 10 = 100$$

$$10^3 = 10 \times 10 \times 10 = 1{,}000$$

In addition, we should recognize that $10^1 = 10$, and $10^0 = 1$. To write a number

---

[1] The length of the meter is now defined as equal to 1,650,763.73 wavelengths of a certain line in the spectrum of the element *krypton* measured in a vacuum.

TABLE 1-1 Metric units of linear measurement

UNIT	ABBREVIATION	METER EQUIVALENT
1 *mega*meter	Mm	1,000,000, or 1×10^6
*1 *kilo*meter	km	1,000, or 1×10^3
1 *hecto*meter	hm	100, or 1×10^2
1 *deka*meter	dam	10, or 1×10^1
*1 meter	m	1, or 1×10^0
1 *deci*meter	dm	0.1, or 1×10^{-1}
*1 *centi*meter	cm	0.01, or 1×10^{-2}
*1 *milli*meter	mm	0.001, or 1×10^{-3}
1 *micro*meter (micron)	μm	0.000001, or 1×10^{-6}
1 *nano*meter (millimicron)	nm	0.000000001, or 1×10^{-9}
1 angstrom	Å	0.0000000001, or 1×10^{-10}

*Most frequently used units.

such as 347 as a power of 10, we recognize that 3.47×10^2 is the same as 3.47×100, and this is the same as 347. In other words, we break the number into two parts: The first part of the number tells the integers that are involved (1, 2, 3, etcetera); and the second part of the number, or the *power of 10,* tells us where to place the decimal point. Other examples are the following:

$$5,280 = 5.28 \times 1,000 = 5.28 \times 10^3$$

$$0.024 = 2.4 \times \tfrac{1}{100} = 2.4 \times 10^{-2}$$

TABLE 1-2 Metric units of volume

UNIT	ABBREVIATION	LITER EQUIVALENT
1 *kilo*liter	kl	1,000 (1×10^3)
1 *hecto*liter	hl	100 (1×10^2)
1 *deka*liter	dal	10 (1×10^1)
*1 liter	l	1 (1×10^0)
1 *deci*liter	dl	0.1 (1×10^{-1})
1 *centi*liter	cl	0.01 (1×10^{-2})
*1 *milli*liter (cubic centimeter)	ml cc	0.001 (1×10^{-3})
1 *micro*liter (lambda)	μl (λ)	0.000001 (1×10^{-6})

*Most frequently used units.

(See Appendixes I and II for further work on exponential notation.)

To help us obtain a better feeling for the size of the metric system units of length, remember that the meter is just a little longer than the English *yard*. Many *yardsticks* are actually made to be 1 meter in length. Also there are approximately $2\frac{1}{2}$ centimeters to 1 inch, and the kilometer is approximately five-eighths of 1 mile. (*Note:* These are *not* exact conversion factors.)

The fundamental unit of *volume* in the metric system is the *liter*. The standard volume that was established as 1 liter was originally intended to occupy a volume of exactly 1,000 cubic centimeters. Although the original standard volume adopted was just slightly larger than was intended (and some texts still mention this), the definition of a liter was changed so that now one liter is equal

TABLE 1-3 Metric units of weight

UNIT	ABBREVIATION	GRAM EQUIVALENT
*1 *kilo*gram	kg	1,000 (1×10^3)
1 *hecto*gram	hg	100 (1×10^2)
1 *deka*gram	dag	10 (1×10^1)
*1 gram	g	1 (1×10^0)
1 *deci*gram	dg	0.1 (1×10^{-1})
1 *centi*gram	cg	0.01 (1×10^{-2})
*1 *milli*gram	mg	0.001 (1×10^{-3})
1 *micro*gram	μg	0.000001 (1×10^{-6})

*Most frequently used units.

to exactly one-thousand cubic centimeters; this makes one milliliter equal to exactly one cubic centimeter.

To help us obtain a feeling for the volume units used here, we should recognize that 1 liter is just slightly larger than 1 quart. The milliliter (or cubic centimeter) is about equal to the volume of 20 drops of water. Table 1-2 lists the volume units used in the metric system.

The fundamental unit of *weight* in the metric system is the *gram*. The gram was established as the weight of one cubic centimeter of pure water measured at a temperature of four degrees Celsius. Also established were larger and smaller units of weight, a few of which are given in Table 1-3. To obtain a feeling for the size of these metric system units of weight, the gram is equal to about one-twenty-eighth of an ounce, and the kilogram is a little more than two pounds. (These comparisons should *not* be used as conversion factors but are intended only to help us develop a feeling for the approximate amounts that we use in the metric system.)

1-5 CONVERSION OF UNITS

A few of the important relationships between the commonly used English system and the metric system are indicated by the equalities shown in Table 1-4. Since both English and metric systems are used today, it is necessary to be able to convert measurements from one system to the other. A method called *dimensional analysis* is used for the conversion from one type of unit to another type of unit. In using this method, the first step is to *write down* the measurement that is originally made; for example, if you measured a piece of wood and found it to be 10 inches long, this would be written down as 10 inches.

The second step is to decide what *conversion factor* is necessary and then use this conversion factor in a calculation to convert from the unit you have to the new unit that is desired. To find the necessary conversion factor you may look in a table such as found in Appendix III of this book or in the "Handbook of Chemistry and Physics."[1] A conversion factor is formed by setting up a fraction equal to 1; this is done by dividing one number by another number *that it is equal to*, including the unit with each number.

Suppose that it is desired to know the length of a 10-inch block of wood expressed in meters. In the table of conversion factors (see Table 1-4 and Appendix III), we see that 1 meter = 39.37 inches. From this information, it is pos-

TABLE 1-4 Some useful equalities

ENGLISH SYSTEM	METRIC SYSTEM
1.00 pound	= 454 grams
1.00 inch	= 2.54 centimeters
1.00 meter	= 39.37 inches
1.00 quart	= 0.946 liter

[1] Published by the Chemical Rubber Co., Cleveland, Ohio.

sible to set up two different conversion factors: 1 meter/39.37 inches and 39.37 inches/1 meter. Both are correct conversion factors, but the first one permits us to carry out the desired conversion because it permits the unit inches to *cancel out*.

$$10 \text{ inches} \times \frac{1 \text{ meter}}{39.37 \text{ inches}} = 0.254 \text{ meter}$$

Note: This leaves the unit meters in the answer.

To illustrate this method further, let us calculate the length in centimeters of a rope that is 2 feet 4 inches long. The first step would be to express the length of the rope in a single English unit—*inches* would be convenient here. Thus

$$2 \text{ feet} \times \frac{12 \text{ inches}}{1 \text{ foot}} = 24 \text{ inches}$$

Then 24 inches + 4 inches = 28 inches (the length of the rope in inches). To convert to meters, we have

$$28 \text{ inches} \times \frac{1 \text{ meter}}{39.37 \text{ inches}} = 0.711 \text{ meter}$$

and to convert to centimeters:

$$0.711 \text{ meter} \times \frac{100 \text{ centimeters}}{1 \text{ meter}} = 71.1 \text{ centimeters}$$

In summary,

$$28 \text{ inches} \times \frac{1 \text{ meter}}{39.37 \text{ inches}} \times \frac{100 \text{ centimeters}}{1 \text{ meter}} = 71.1 \text{ centimeters}$$

Note that the same arithmetical operations are carried out on the *units* as are carried out on the *numbers*. This is the method of *dimensional analysis*. When this method is mastered, simple though it is, it can be a very powerful tool in problem solving.

EXAMPLE   The engine volume of a Ferrari sports car is 4.8 liters. Calculate the volume in cubic inches.

We first calculate the volume in cubic centimeters:

$$4.8 \text{ liters} \times \frac{1{,}000 \text{ cc}}{1 \text{ liter}} = 4.8 \times 10^3 \text{ cc}$$

Then we calculate the number of cubic centimeters in 1 in.^{3}:

$$1 \text{ in.}^3 = 1 \text{ in.} \times 1 \text{ in.} \times 1 \text{ in.} = 2.54 \text{ cm} \times 2.54 \text{ cm} \times 2.54 \text{ cm}$$
$$= 16.4 \text{ cc}$$

Then we obtain

$$4.8 \times 10^3 \, \cancel{cc} \times \frac{1 \text{ in.}^3}{16.4 \, \cancel{cc}} = 0.292 \times 10^3 \text{ in.}^3$$

$$\text{or} \quad 2.92 \times 10^2 \text{ in.}^3 \quad \text{or} \quad 292 \text{ in.}^3$$

The standard 30-caliber rifle has an inside barrel diameter of 0.30 in. **EXAMPLE**
Calculate the inside diameter of the barrel in millimeters.

From Table 1-4, 2.54 cm/l in. is a conversion factor from inches to centimeters. Therefore,

$$0.300 \, \cancel{\text{in.}} \times \frac{2.54 \text{ cm}}{1 \, \cancel{\text{in.}}} = 0.762 \text{ cm}$$

Then

$$0.762 \, \cancel{\text{cm}} \times \frac{10 \text{ mm}}{1 \, \cancel{\text{cm}}} = 7.62 \text{ mm}$$

To summarize the calculation:

$$0.300 \, \cancel{\text{in.}} \times \frac{2.54 \, \cancel{\text{cm}}}{1 \, \cancel{\text{in.}}} \times \frac{10 \text{ mm}}{1 \, \cancel{\text{cm}}} = 7.62 \text{ mm}$$

To summarize this method:

1. Write down the original measurement.
2. Look up any necessary conversion factors.
3. Carry out indicated multiplication and division on the units and also on the corresponding numbers.

1-6 TEMPERATURE CONVERSION

Normally either the Fahrenheit or Celsius temperature scale is used for the measurement of temperature. The Celsius temperature scale was established by arbitrarily defining the freezing point of pure water as 0 and the boiling point of pure water at atmospheric pressure as 100, thus establishing the size of the degree since there are 100 equal divisions between 0 and 100. The Fahrenheit temperature scale was established by taking the freezing point of a saturated ice-salt mixture as 0 degrees and body temperature as 100 degrees.

Conversions between the Fahrenheit and Celsius temperature scales may be made by comparing the two as in Fig. 1-1. We observe that 100 degrees on the Celsius scale covers the same temperature interval as 180 degrees on the Fahrenheit scale.

From the figure, we observe that an algebraic relationship may be set up:

$$\frac{°C}{100} = \frac{°F - 32}{180}$$

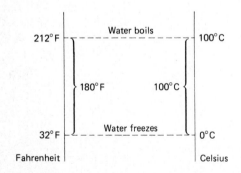

FIGURE 1-1 Comparison of Fahrenheit and Celsius temperature scales

We observe that it is necessary to subtract 32 degrees from the reading on the Fahrenheit scale in order to reach the point on the Fahrenheit scale that corresponds to 0 degrees Celsius. To solve this equation for degrees Celsius,

$$\frac{°C \times 20}{100} = \frac{(°F - 32) \times 20}{180}$$ multiplying both sides of the equation by 20

$$\frac{°C}{5} = \frac{(°F - 32)}{9}$$ canceling 20 on each side of the equation

$$°C = \frac{5 \, (°F - 32)}{9}$$ multiplying both sides of the equation by 5

$$= \tfrac{5}{9} \, (°F - 32) = 0.555 \, (°F - 32)$$

Derive the equation $°F = \tfrac{9}{5} \, °C + 32$ from the equation **EXAMPLE**

$$\frac{°C}{100} = \frac{°F - 32}{180}$$

First multiply both sides of the equation by 180:

$$180 \times \frac{°C}{100} = °F - 32$$

Then simplify the fraction on the left-hand side of the equation:

$$\tfrac{9}{5} \times °C = °F - 32$$

Now add 32 to each side of the equation:

$$\tfrac{9}{5} \times °C + 32 = °F$$

$$°F = \tfrac{9}{5} °C + 32$$

It is suggested that the equation

$$\frac{°C}{100} = \frac{°F - 32}{180}$$

be learned as the most useful equation for making temperature conversions between Fahrenheit and Celsius temperature scales. The relationships shown in Fig. 1-1 will help you remember it.

EXAMPLE   Calculate the reading on the Celsius scale if the reading on the Fahrenheit scale is 86°F.

We first recall the equation for temperature conversion discussed above, and, substituting 86°, we have

$$\frac{°C}{100} = \frac{86-32}{180}$$

$$= \frac{54}{180} \quad \text{subtracting 32 from 86}$$

$$\frac{°C}{100} \times 100 = \frac{54}{180} \times 100 \quad \text{multiplying both sides of the equation by 100}$$

$$°C = 54 \times \tfrac{5}{9} \quad \text{canceling 100 on the left side of the equation and 20 on the right side}$$

$$= 6 \times 5 \quad \text{canceling 9 on the right side of the equation}$$

$$= 30$$

In a similar manner, we can make a conversion from a reading on the Celsius scale to a reading on the Fahrenheit scale by means of the same equation but substituting for degrees Celsius.

EXAMPLE

The reading on a Celsius thermometer is 45°C. Calculate the corresponding reading on a Fahrenheit thermometer.

$$\frac{45}{100} = \frac{°F - 32}{180} \quad \text{substituting 45 for °C}$$

$$\frac{45 \times 180}{100} = \frac{(°F - 32)\,180}{180} \quad \text{multiplying both sides of the equation by 180}$$

$$\frac{45 \times 9}{5} = °F - 32$$ canceling 20 on the left side of the equation and 180 on the right side

$$9 \times 9 = °F - 32$$ canceling 5 on the left side of the equation

$$81 + 32 = °F = 113°$$ multiplying 9×9 and adding 32 to both sides

1-7 DENSITY

Another concept that should be introduced at this time is the concept of *density*. The density of a substance may be defined as the mass (weight) of the substance divided by its volume. If we think about this concept a bit, we shall see there are many different units that might be used for the expression of density. We might express density as pounds/gallon, ounces/pint, pounds/cubic foot or in various other ways. In chemistry, we most frequently express density as *grams/milliliter*. We remember that the gram was defined as the weight of one milliliter of water at exactly four degrees Celsius. Thus, the density of water is exactly 1 gram/milliliter at 4 degrees Celsius and will be nearly 1 gram/milliliter at any temperature below its boiling point.

A substance other than water will have its own characteristic density. Carbon tetrachloride, for example, has a density of 1.59 grams/milliliter. This means that 1 milliliter of carbon tetrachloride weighs 1.59 grams. Carbon in the form of diamond has a density of 3.51 grams/milliliter, and iron has a density of 7.86 grams/milliliter. These densities have been experimentally measured.

There is no theory that will permit us to predict the density of a substance without making some measurements.

In the density formula

$$\text{Density} = \frac{\text{weight}}{\text{volume}}$$

there are three factors. If any two of these factors are known, the other one may be calculated. Thus, if a certain sample weighs 12 grams and occupies a volume of 9 milliliters, the density of the substance would be

$$\text{Density} = \frac{12 \text{ grams}}{9 \text{ milliliters}} = 1.33 \text{ grams/milliliter}$$

EXAMPLE The density (D) of copper is 8.92 g/ml. Calculate the weight of 15.0 ml of copper.

$$D = \frac{\text{wt}}{\text{vol}}$$ the formula for density is written

$$8.92 \text{ g/ml} = \frac{\text{wt}}{15.0 \text{ ml}}$$ 8.92 g/ml is substituted for D, and 15.0 ml is substituted for vol

$$8.92 \text{ g/ml} \times 15.0 \text{ ml} = \frac{\text{wt} \times 15.0 \text{ ml}}{15.0 \text{ ml}}$$ multiply both sides of the equation by 15.0 ml

8.92g × 15.0 = wt = 133.8g   or about 134 g

The number of figures that should be carried in the answer to a problem will depend upon the number of figures, or, more exactly, the number of *significant figures,* that are given in the data. In this book, we shall not discuss significant figures; instead we shall establish the general rule that slide-rule accuracy will be accepted for all problems. This means that *three digits* will be considered satisfactory.

As another example dealing with the density formula, let us calculate the volume of a substance when the density and the weight are known.

EXAMPLE Lead is known to have a density of 11.34 g/ml. Calculate the volume occupied by 40.0g of lead.

$$D = \frac{wt}{vol}$$ write down the density formula

$$11.34 \text{ g/ml} = \frac{40.0g}{vol}$$ substitute the value of the density and the weight

$$11.34 \text{ g/ml} \times vol = \frac{40.0g \times \cancel{vol}}{\cancel{vol}}$$ multiply both sides of the equation by volume

$$Vol = \frac{40.0g}{11.34 \cancel{g}/ml}$$ divide both sides of the equation by 11.34 g/ml

$$= 3.53 \text{ ml}$$

Notice the units in the last step of this calculation: grams divided by grams/milliliter. This is actually a complex fraction and needs to be handled as such. To work this through in detail:

$$\frac{\text{Grams}/1}{\text{Grams}/\text{milliliter}} = \frac{\text{grams}}{1} \div \frac{\text{grams}}{\text{milliliter}} = \frac{\cancel{\text{grams}}}{1} \times \frac{\text{milliliter}}{\cancel{\text{grams}}} = \text{milliliters}$$

Thus we see that the answer is in milliliters.

With a little practice in the handling of units, we shall find that they serve as an excellent guide in the solution of many problems. For this reason, the usefulness and importance of carefully following units in calculations cannot be overemphasized.

1-8 SPECIFIC GRAVITY

The term *specific gravity* may be encountered in problems dealing with weight and volume. Specific gravity is defined as the *ratio* of the weight of a certain volume of a substance to the weight of an equal volume of some reference substance, usually taken as water. For example: 1 milliliter of lead weighs 11.34 grams; 1 milliliter of water weighs 1 gram. The *ratio* of the weight of 1 milliliter of lead to 1 milliliter of water is

$$\frac{11.34 \; \cancel{\text{grams}}}{1.00 \; \cancel{\text{grams}}} = 11.34$$

Note that specific gravity is a dimensionless number. In this case we observe that the numerical value of the specific gravity is the same as the numerical value of the density. The numerical value of specific gravity will always be the same as the numerical value of density when water is taken as the reference standard and metric units are used for density (grams/milliliter).

If other units are used, the numerical value of specific gravity is quite different from the numerical value of density. For example, 1 cubic foot of lead weighs 710 pounds, or its density may be expressed as 710 pounds/cubic foot. Since 1 cubic foot of water weighs 62.4 pounds, the specific gravity of lead is

$$\frac{710 \text{ pounds/cubic foot}}{62.4 \text{ pounds/cubic foot}} = 11.34$$

Comparing 710 pounds/cubic foot with 11.34, we see that, using these units, the specific gravity is very different from the density.

GLOSSARY

Celsius temperature: The temperature expressed using the Celsius temperature scale; $°C = \frac{5}{9}(°F - 32)$

Chemistry: The study of the structure of matter, the changes that matter will undergo, and the associated energy changes

Compound: A substance that is composed of two or more elements chemically combined in a definite proportion by weight

Density: Ratio of mass to volume; $D = \text{wt/vol}$

Dimensional analysis: A method used for the conversion from one type of unit to another (i.e., from English system units to metric system units or vice versa)

Element: A substance that cannot be decomposed into simpler substances by ordinary processes

Gram: The fundamental unit of weight in the metric system (about $\frac{1}{28}$ of an ounce)

Heterogeneous material: One that is nonuniform throughout

Homogeneous material: One that is uniform throughout

Hypothesis: An imaginary picture suggested to explain observations (see Theory)

Law: A generalized statement that summarizes a group of facts

Liter: The fundamental unit of volume in the metric system (about 1.05 qt)

Matter: Any real substance which is attracted by gravity

Meter: The fundamental unit of length in the metric system (about 39 in.)

Metric system: A system of measurements using the meter, the liter, and the gram as fundamental units

Mixture: Material that can be separated into its different parts by mechanical processes

Properties: Characteristics by which a substance may be identified

Scientific method: A method of analyzing a problem (see pages 6 to 8)

Scientific Notation: A method of writing numbers as powers of 10

Specific gravity: The ratio of the weight of a certain volume of a substance to the weight of an equal volume of some reference substance

Substance: Usually taken to mean a pure, or single kind, of material

Theory: A model based on familiar concepts which explains observations

SELF TEST 1 In the metric system, the gram is a unit of ___wt___; the liter is a unit of ___volume___; the meter is a unit of ___length___.

2. Write 2,437 using scientific notation. 2.437×10^3

3. Convert 150 mg into grams. 0.150 gms.

4. Convert 50.0 liters into milliliters. 5.00×10^4 ml.

5. Convert 0.357 mm into meters. 35.7×10^{-2}

6. Express a temperature of 200° F in degrees Celsius.

7. The units of density are usually _____.

8. A liquid has a volume of 20.0 ml and weighs 35.0 g. What is the density of this liquid?

ANSWERS

1. Mass (or weight); volume; length
2. 2.437×10^3
3. 0.150 g
4. 50.0×10^3 ml, or 5.00×10^4 ml
5. 0.357×10^{-3} m, or 3.57×10^{-4} m
6. 93.4°C
7. grams/milliliter
8. 1.75 g/ml

EXERCISES

1. What is chemistry? ("A required course" is not an acceptable answer.)

2. What is the difference between a *law* and a *theory*, as the terms are used in chemistry?

3. Define the terms
   (a) Properties
   (b) Element
   (c) Compound
   (d) Mixture

4. What are the advantages of the metric system?

5. What would be the unit on the answer if
   (a) A distance in feet is multiplied by inches/foot, then by meters/inch, and then by millimeters/meter
   (b) A volume in quarts is multiplied by liters/quart, then by milliliters/liter, and then by grams/milliliter
   (c) A weight in ounces is multiplied by pounds/ounce, then by grams/pound, and the result is divided by grams/milliliter

6. To convert inches to centimeters, you must multiply by the conversion factor (a) 2.54 cm/in. To convert meters to millimeters, you must multiply by the conversion factor (b) _____. To convert meters to centimeters, you must multiply by the conversion factor (c)

_____. To convert centimeters to meters, you must multiply by the conversion factor (d) _____. To convert kilometers to centimeters, you must multiply by the conversion factor (e) _____. To convert quarts to liters, you must multiply by the conversion factor (f) _____. To convert from liters to quarts, you must multiply by the conversion factor (g) _____. To convert liters to milliliters, you must multiply by the conversion factor (h) _____. To convert grams to kilograms, you must multiply by the conversion factor (i) _____. To convert kilograms to milligrams, you must multiply by the conversion factor (j) _____.

7   A basketball player is 7 ft 1 in. tall. Express his height in
    (a) Meters
    (b) Centimeters
    (c) Angstroms

8   A football pass play gained 25 yd. Calculate
    (a) How many feet were gained
    (b) How many millimeters were gained

9   How many yards are there in 1,000 m?

10 Calculate the number of liters displacement in a 318-in.^{3} automobile engine.

11. A metric ton is 1,000 kg. How many pounds are there in a metric ton?

12. Calculate
    (a) The number of pints that are contained in 1 liter
    (b) The number of cubic centimeters in 6.4 fl oz of a baby's formula (16.0 fl oz = 1 pint).

13. What would be the weight in grams of a 4 oz sample of sodium chloride?

14. A piece of steak weighs 2 lb 3 oz. Calculate its weight in
    (a) Grams
    (b) Milligrams
    (c) Micrograms

15. Liquid nitrogen boils at $-196°C$. What is the Fahrenheit temperature?

16. The reading on a Fahrenheit thermometer is 72°F. What would be the reading on a Celsius thermometer that is placed alongside this one?

17. (a) What would be the reading on a Fahrenheit thermometer if the reading on a Celsius thermometer is 32°C?
    (b) A meat thermometer inserted in a roast reads 170°F when the roast is done. What would be its temperature on the Celsius scale?

18. Dry Ice has a temperature of −78°C. What is the Fahrenheit temperature?

19. The temperature on the surface of the sun is about 6000°C. What is the Fahrenheit temperature?

20. Calculate the density of iron if a 15.0-ml sample weighs 117.0 g.

21. What would be the volume of a piece of gold that weighs 35 g? (The density of gold is 19.3 g/ml.)

22. Osmium, the densest element known, has a density of 22.6 g/ml. How much would 1 liter of osmium weigh?

23. What would be the weight of 1 qt of gold? (Density = 19.3 g/ml.)

24. What would be the weight in pounds of 3 liters of lead? (Density = 11.3 g/ml.)

25. Calculate the temperature at which you would find twice the numerical reading on the Fahrenheit scale as observed on the Celsius scale. [*Hint:* Set up an algebraic equation, making use of the information in the problem and the formula °C/100 = (°F − 32)/180.]

26. The line left by a pencil is 0.35 mm wide. Calculate the width in
    (a) Microns
    (b) Millimicrons
    (c) Angstroms

27. A single bacterium of a common type measures $\frac{1}{25,400}$ in. in length. Convert this length to microns.

28. A football field is 100 yd long. How many meters is this?

29. A grain of sand weighs 2.1mg. What is the weight in
    (a) Micrograms
    (b) Ounces

30. The density of silver is 10.5 g/ml. Calculate the weight in ounces of a silver coin that has a volume of 0.25 ml.

31. If a silver coin weighs 12.29 g, what volume of silver does it contain (expressed in cubic centimeters)?

32. Calculate the density of copper in grams/milliliter if 1 ft^3 weighs 556 lb.

33. The density of lead is 11.3g/ml. Calculate the density in lb/ft^3 if the density of water is 62.4 lb/ft^3.

34 The specific gravity of copper is 8.92. The density of water is 62.4 lb/ft^3.
   (a) Calculate the weight in pounds of 3 ft^3 of copper.
   (b) Calculate the volume in liters of a 500-lb bar of blister copper.

35 A spacecraft is to be plated with gold. If the total surface area is 10 m^2 and the plate thickness is to be 1.2 μm, calculate the number of grams of gold required.

36 In an average year the people in the Los Angeles area use about 170 billion gal of water. Calculate the amount of water used in
   (a) Grams
   (b) Kilograms
   (c) Pounds

THE ELEMENTS

CHAPTER TOPICS

Before learning anything else about chemistry, you will need to know

How various elements can be classified as *metals*, *nonmetals*, or *inert elements* by their properties

The *symbols* for common elements

What the *periodic table* is

How the periodic table is organized into groups and periods of elements

How the periodic table can be used to predict the properties of an element

2-1 INTRODUCTION

The ancient Greeks believed that the world and everything in it was composed of four fundamental substances, or elements: earth, air, fire, and water. This idea persisted for many centuries, and it was not until about the time of the Renaissance that men began to recognize that earth, air, and water represent different states of matter. Today we call these states of matter *solid, gas,* and *liquid*. In general, any substance may exist in any one of these states under suitable temperature and pressure conditions. We now know that fire is not a substance, or element, at all but involves a chemical change with an accompanying energy change.

At about the beginning of the sixteenth century, men began to recognize certain materials as *fundamental substances*, that is, substances which could not be changed into simpler substances by chemical processes. From that time to the present there have been many such fundamental substances isolated from naturally occurring materials. We call these fundamental substances *elements*. What, then, are these elements of which the earth is composed?

2-2 BROAD CLASSIFICATIONS OF THE ELEMENTS

There are three types of elements: *metals, nonmetals*, and *inert gases*. The first type of elements we shall discuss are the *metals*. Undoubtedly you are familiar with many of the metallic elements, such as copper, iron, lead, zinc, silver, and gold. These elements have certain common characteristics: They all have the so-called metallic sheen, or luster; they will all conduct an electric current; and they are all good conductors of heat. It is interesting to note that about 80 percent of the total number of known elements are metals.

The second type of elements we shall discuss are the *nonmetals*. Some familiar examples of nonmetals are sulfur, oxygen, nitrogen, and carbon. These elements have little, if any, metallic sheen, or luster, and do not conduct electric current or heat as well as metals. Nonmetals have other properties in common also, but these statements suffice to show that the nonmetallic elements are different from the metals in some fundamental way.

The third type of element, the *inert gases*, are fundamentally different from either the metals or the nonmetals; they are all gases at room temperature. But they have a more important characteristic: they have a very slight tendency to combine with other elements. This means that the inert gases will not form compounds easily. Prior to 1962 it was believed that the inert gases would not form *any* chemical compounds; but since that time, a few compounds have been formed which include inert elements.

At about the turn of the century (1900) scientists began to investigate radioactive phenomena. Progress in this field led to reports (beginning in 1937) of the formation of new, *man-made* elements, amazing as this may have seemed to chemists of earlier times. All the man-made elements are solids, and with the exception of astatine all of them are metals. All these man-made elements decompose into other elements and are therefore said to be *radioactive*; some of them decompose much more rapidly than others. It is assumed that these elements were probably present when the earth was formed, but that they have decomposed to such an extent that they can no longer be detected in naturally occurring materials. One of them, *plutonium*, is a very important element in the atomic energy program because it is one of the materials that is used in the production of atomic energy and atomic bombs.

2-3   NAMES AND SYMBOLS OF THE ELEMENTS

Each element has a *name* and a *symbol*. Before we can proceed further with our study of chemistry, we shall need to learn the names and symbols of the common elements. The *symbol* of an element is an *abbreviation* of the name of that element, and the abbreviation is formed in a definite way. The symbol is formed by taking the first letter of the name of the element as a capital letter; for some elements, that single capital letter is the complete symbol. Since there are many more elements than letters in the alphabet, it is necessary to add a second letter to the symbols of some elements in order to distinguish among them. For

TABLE 2-1 Elements with symbols derived from foreign names

ENGLISH NAME	FOREIGN NAME	SYMBOL
Antimony	*stibium*	Sb
Copper	*cuprum*	Cu
Gold	*aurum*	Au
Iron	*ferrum*	Fe
Lead	*plumbum*	Pb
Mercury	*hydrargyrum*	Hg
Potassium	*kalium*	K
Silver	*argentum*	Ag
Sodium	*natrium*	Na
Tin	*stannum*	Sn
Tungsten	*wolfram*	W

example, helium, element number 2, has the symbol He. If a second letter is present in a symbol, it is a lowercase letter; this second letter is a significant letter in the name. The symbol of an element is used to represent one atom of that element. (We shall discuss the atom in Chap. 3.)

If all elements had symbols that were abbreviations of the English names of the elements, it would be easy (for English-speaking students) to learn them. However, some elements have symbols that are abbreviations of foreign names. These elements, with their foreign names and symbols, are listed in Table 2-1; note that all are derived from Latin words (with the exception of tungsten, which is derived from the German word *wolfram*). It is suggested that the student learn the English name, the foreign name, and the symbol of each ele-

TABLE 2-2 Names and symbols of other naturally occurring elements

NAME	SYMBOL	NAME	SYMBOL
Actinium	Ac	Holmium	Ho
*Aluminum	Al	*Hydrogen	H
*Argon	Ar	Indium	In
*Arsenic	As	*Iodine	I
*Barium	Ba	Iridium	Ir
*Beryllium	Be	Krypton	Kr
Bismuth	Bi	Lanthanum	La
*Boron	B	*Lithium	Li
*Bromine	Br	Lutetium	Lu
*Cadmium	Cd	*Magnesium	Mg
*Calcium	Ca	*Manganese	Mn
*Carbon	C	Molybdenum	Mo
Cerium	Ce	Neodymium	Nd
Cesium	Cs	*Neon	Ne
*Chlorine	Cl	*Nickel	Ni
*Chromium	Cr	Niobium	Nb
*Cobalt	Co	*Nitrogen	N
Dysprosium	Dy	Osmium	Os
Erbium	Er	*Oxygen	O
Europium	Eu	Palladium	Pd
*Fluorine	F	*Phosphorus	P
Francium	Fr	Platinum	Pt
Gadolinium	Gd	Polonium	Po
Gallium	Ga	Praseodymium	Pr
Germanium	Ge	Protactinium	Pa
Hafnium	Hf	Radium	Ra
*Helium	He	Radon	Rn

TABLE 2-2 Names and symbols (*continued*)

NAME	SYMBOL	NAME	SYMBOL
Rhenium	Re	Terbium	Tb
Rhodium	Rh	Thallium	Tl
Rubidium	Rb	Thorium	Th
Ruthenium	Ru	Thulium	Tm
Samarium	Sm	Titanium	Ti
Scandium	Sc	*Uranium	U
Selenium	Se	Vanadium	V
*Silicon	Si	Xenon	Xe
Strontium	Sr	Ytterbium	Yb
*Sulfur	S	Yttrium	Y
Tantalum	Ta	*Zinc	Zn
Tellurium	Te	Zirconium	Zr

*Most common elements.

TABLE 2-3 Names and symbols of man-made elements

NAME	SYMBOL	NAME	SYMBOL
Americium	Am	Lawrencium	Lr
Astatine	At	Mendelevium	Md
Berkelium	Bk	Neptunium	Np
Californium	Cf	Nobelium	No
Curium	Cm	Plutonium	Pu
Einsteinium	Es	Promethium	Pm
Fermium	Fm	Technetium	Tc

ment listed in Table 2-1. Most of these are common elements that are encountered frequently; by learning these symbols that *are not* related to English names we might more easily remember symbols of elements that *are* abbreviations of English names.

Table 2-2 lists the names and symbols of other naturally occurring elements; and Table 2-3 lists the names and symbols of man-made elements. The more common elements are the only ones that will be discussed in this text.

2-4 SOME COMMON ELEMENTS

While we are discussing the elements, we should consider those elements that are most abundant on the surface of the earth. Let us first consider the element *oxygen*, which is the most abundant of all the elements. If we consider the total weight of material available to us on the earth's surface, the element oxygen represents about 50 percent of the total weight (49.2 percent). That is, if we combine the weight of the atmosphere, the water (including the oceans), and all the land (down to a depth of about 10 miles), nearly one-half of the weight of this material would be oxygen. (However, most of this oxygen is not in the form of the *free* element; most of it is combined with other elements in the form of *compounds*. Water, for example, is 88.8% oxygen by weight.)

Pure oxygen is a gas at room temperature. It is colorless, odorless, and will not burn. If a substance that is already burning is placed in pure oxygen, it will burn more rapidly. Some substances, such as iron, will burn readily in pure oxygen but not in air. These characteristics are some of the properties of oxygen.

Another abundant element on the earth is *silicon*. Compounds containing silicon make up most of the rock, sand, and other solid parts of the earth. In these solids, silicon is combined with oxygen and metals such as aluminum, iron, and calcium. About 25 percent (25.17 percent) of the total weight of the crust of the earth is silicon.

The other elements that make up an appreciable portion of the weight of the earth are aluminum (7.5 percent), iron (4.7 percent), calcium (3.3 percent), sodium (2.6 percent), potassium (2.4 percent), and magnesium (1.9 percent). Other elements are present in the earth's crust in relatively small amounts, calculated on a weight percent basis.

2-5 THE PERIODIC TABLE

As chemists discovered the many different elements that we have mentioned, they were continually watching for similarities among them. It is obvious that it would require a great deal of effort to attempt to remember the properties of each element on an individual basis. It would be a tremendous undertaking to attempt to remember the chemical compounds that each element will form plus the many different reactions that each element will undergo. Fortunately, it has been found that there are definite similarities between certain *groups of elements*. If we know the properties and reactions of one of the elements in a certain group, we may predict with reasonable accuracy the properties and reactions of other elements in that group. Also, if we know what compounds will be formed by one of the elements in a group, frequently we can predict the compounds that will be formed by the other elements in that group. When the

IA																	Inert gases
1 H 1.01	IIA											IIIA	IVA	VA	VIA	VIIA	2 He 4.00
3 Li 6.94	4 Be 9.01											5 B 10.8	6 C 12.0	7 N 14.0	8 O 16.0	9 F 19.0	10 Ne 20.2
11 Na 23.0	12 Mg 24.3											13 Al 27.0	14 Si 28.1	15 P 31.0	16 S 32.1	17 Cl 35.5	18 Ar 39.9
19 K 39.1	20 Ca 40.1	21 Sc 45.0	22 Ti 47.9	23 V 50.9	24 Cr 52.0	25 Mn 54.9	26 Fe 55.8	27 Co 58.9	28 Ni 58.7	29 Cu 63.5	30 Zn 65.4	31 Ga 69.7	32 Ge 72.6	33 As 74.9	34 Se 79.0	35 Br 79.9	36 Kr 83.8
37 Rb 85.5	38 Sr 87.6	39 Y 88.9	40 Zr 91.2	41 Nb 92.9	42 Mo 95.9	43 Tc (99)	44 Ru 101.0	45 Rh 102.9	46 Pd 106.4	47 Ag 107.9	48 Cd 112.4	49 In 114.8	50 Sn 118.7	51 Sb 121.8	52 Te 127.6	53 I 126.9	54 Xe 131.3
55 Cs 132.9	56 Ba 137.3	57 La 138.9	72 Hf 178.5	73 Ta 180.9	74 W 183.9	75 Re 186.2	76 Os 190.2	77 Ir 192.2	78 Pt 195.0	79 Au 197.0	80 Hg 200.6	81 Tl 204.4	82 Pb 207.2	83 Bi 209.0	84 Po (210)	85 At (210)	86 Rn (222)
87 Fr (223)	88 Ra (226)	89 Ac (227)															

Lanthanum series

58 Ce 140.1	59 Pr 140.9	60 Nd 144.2	61 Pm (147)	62 Sm 150.4	63 Eu 152.0	64 Gd 157.3	65 Tb 158.9	66 Dy 162.5	67 Ho 164.9	68 Er 167.3	69 Tm 168.9	70 Yb 173.0	71 Lu 175.0

Actinium series

90 Th 232.0	91 Pa (231)	92 U 238.0	93 Np (237)	94 Pu (242)	95 Am (243)	96 Cm (247)	97 Bk (247)	98 Cf (249)	99 Es (254)	100 Fm (253)	101 Md (256)	102 No (253)	103 Lw (257)

TABLE 2-4 Periodic chart of the elements

groups of similar elements are arranged in a table, we have what is called the *periodic table of the elements*, Table 2-4.

The elements are represented in the periodic table by their symbols. Notice that a number is placed above the symbol of the element in the table. This is its *atomic number*. Notice also that a number is placed below the symbol of the element. This is its *atomic weight*. (We shall discuss the meaning of these numbers in Chap. 3.)

The periodic table is one of the most important tools for the chemist. It helps him to remember and to correlate a great deal of information about the different elements. It has been used in research in the past and is so used today. For example, in 1869, when the Russian chemist Dmitri Mendeleev first arranged the elements in approximately the order that is used in the modern periodic table, he correctly predicted the existence of six elements that had never been observed up to that time. He recognized that it would be necessary for these elements to exist in order to fill gaps that were present in his table. Other chemists made use of his predictions in searching for these elements.

Many of the predictions made by Mendeleev were amazingly accurate. He correctly predicted the existence of the element *germanium* (he called it *ekasilicon* because he foresaw that it would be very similar to the element silicon); he accurately predicted the properties that were actually observed when the element germanium was isolated. He also predicted the existence and the approximate properties of scandium, gallium, technetium, rhenium, and protactinium. Today the periodic table is used by men of research in much the same way to predict the properties of the new elements that we are attempting to create.

The most important use of the periodic table to the student is as an aid to memory. There are many facts that can be correlated and more easily remembered on the basis of the periodic relationship. As an example of the value of the periodic table, let us consider the group of elements called *group IA*, containing lithium, sodium, potassium, rubidium, cesium, and francium. We are all familiar with the compound sodium chloride (ordinary table salt). We know that it is a white crystalline solid and that it is a stable compound; and it is easy to learn by experiment that it will conduct an electric current if it is dissolved in water. From what we know about sodium chloride, we can correctly predict that each of the other metals in this group of elements will form a stable compound with the element chlorine and that each of these compounds will be a white crystalline solid. Each of these compounds will be moderately soluble in water, as is sodium chloride, and will conduct an electric current when it is in a water solution. Many other properties of the compounds of these metals can be predicted by careful study of the compounds of any one of the metals.

Groups of elements having similar properties are listed in *vertical columns* in the periodic table (Table 2-4). The alkali metals lithium, sodium, potassium, rubidium, cesium, and francium, for example, are located in the same vertical column in the periodic table (group IA, discussed above). It should be emphasized that these elements, although similar, are not *identical*. There are definite differences among the elements in a group and among their compounds. For example, the melting point of sodium chloride is 801°C and the melting point of potassium chloride is 776°C; these are in the same general temperature range, but they are not *identical* temperatures.

At this point, we have presented only a small amount of the information that can be obtained from a study of the periodic table. As we continue with our study of chemistry, we shall often refer back to the periodic table of the elements (Table 2-4), gradually building a more complete understanding of it.

GLOSSARY

Alkali metals: The elements in group IA of the periodic table (Table 2-4), except for H

Group of elements: A vertical column of elements in the periodic table (Table 2-4)

Inert gas: A chemically inert element, located in far right-hand group in the periodic table (does not react)

Metallic element: An element that shows the properties of a metal (see page 37)

Nonmetallic element: An element that shows properties of a nonmetal (see page 37)

Period of elements: A horizontal row of elements on the periodic table (Table 2-4)

Periodic table: An orderly arrangement of the symbols of the elements (Table 2-4)

Symbol of an element: An abbreviation for the name of that element (see page 38)

SELF TEST

1. Gold is an example of a(n) _____ element; oxygen is an example of a(n) _____ element; and helium is a(n) _____ element.

2. Without referring to the text, write the symbol for
   (a) Neon
   (b) Aluminum
   (c) Chlorine
   (d) Sodium
   (e) Potassium

3. Without referring to the text, write the name of the elements
   (a) Br
   (b) S
   (c) P
   (d) Li
   (e) I

4. A vertical column of elements on the periodic table is called a _____ , whereas a horizontal row of elements is called a _____ .

ANSWERS

1. Metallic; nonmetallic; inert
2. (a) Ne; (b) Al; (c) Cl; (d) Na; (e) K
3. (a) Bromine; (b) sulfur; (c) phosphorus; (d) lithium; (e) iodine
4. Group, period

EXERCISES

1. What is an element?

2. What were the ideas of the ancient Greeks about the fundamental substances of which all matter is composed?

3. What is the symbol of an element?

4. There are 11 elements that have symbols that are abbreviations of the Latin or German names of the elements. Write the English name, the Latin (or German) name, and the symbol of each of these elements.

5. Write the names of the elements that are asterisked in Table 2-2, and then write their symbols from memory.

6. The following exercise is designed to help you learn the symbols of the most common elements. Supply the missing symbol for each element. You should be able to do the problem without using your text.
    Aluminum metal is used extensively in beverage containers; the

symbol for aluminum is (a) _____ . Argon (meaning, literally, *lazy one*) is sometimes used to fill light bulbs; the symbol for argon is (b) _____ . The element arsenic is used in the manufacture of some insecticides and rodent poisons; the symbol for arsenic is (c) _____ . Boron, found in borax, has the symbol (d) _____ . One major use of the element bromine is in the synthesis of ethylene dibromide, which is a gasoline additive; the symbol for bromine is (e) _____ . Animal bones are composed of complex substances containing calcium whose symbol is (f) _____ . Lithium carbonate has been found to be effective in the treatment of certain mental disorders; the symbol for lithium is (g) _____ . Chlorophyll is a rather complex substance containing magnesium, whose symbol is (h) _____ . Neon signs contain the element whose symbol is (i) _____ . Steel is often coated with a protective layer of nickel metal, whose symbol is (j) _____ . The atmosphere is about 21% (by volume) oxygen, whose symbol is (k) _____ . Phosphorus, used extensively in the manufacture of fertilizers, has the symbol (l) _____ . Antimony, whose symbol is (m) _____ , is used in the manufacture of linotype metal. Fluorine, whose symbol is (n) _____ , is used in the manufacture of Teflon. Table salt is composed of the elements sodium and chlorine, whose symbols are (o) _____ and (p) _____ . Sulfur, whose symbol is (q) _____ , is used in manufacturing detergents. Traces of the element mercury, whose symbol is (r) _____ , have been found as a contaminant of fish meat. Tungsten is a metal with a very high melting point and is used as light-bulb filaments; the symbol of tungsten is (s) _____ . The element potassium, whose symbol is (t) _____ , is found in potash. Some toothpaste tubes are made out of lead metal whose symbol is (u) _____ .

7   It is believed that the 14 new elements that men have created may have existed naturally upon the earth at one time but no longer do. Why?

8   Why is the single element oxygen so important in the study of chemistry?

9   The melting point of lithium is 186°C, the melting point of sodium is 97.5°C, and the melting point of potassium is 62.3°C. Predict the melting point of rubidium.

10 The boiling points of fluorine, chlorine, bromine, and iodine are −188, −34.6, 58.78, and 184.35°C, respectively. Predict the boiling point of man-made astatine (At).

11 What fundamental differences exist between the metals, the nonmetals, and the inert elements?

12 What is the periodic table?

13 Why is the periodic table important to a student of chemistry?

14 What use might be made of the periodic table in doing chemical research?

15 Would you expect the compound potassium chloride to be exactly like the compound sodium chloride?

16 What type of prediction did Mendeleev make from the periodic table?

17 What are the elements in group IA in the periodic table?

18 Name the elements that have the following symbols: Mg, Mn, Na, K, Cl, Au, Br, Be, He, W, Fe, Ag, Pb, S, and P.

19 Calculate the density of osmium from the following data: 58.0 g occupies 2.58 ml.

THE ATOM

THE ATOM

CHAPTER TOPICS

When you have completed this chapter, you should have an understanding of

The development of modern *atomic theory*

Dalton's atomic theory

Brownian motion

The *cathode ray tube*

The forces of attraction and repulsion existing between charged particles

The number and kind of particles in an *atom*

The *nuclear model* of the atom

The development of *Bohr's theory* of atomic structure

The modification of Bohr's theory to the present-day theory of atomic structure

Isotopes

Atomic weight and the periodicity of properties of the elements

3-1 BRIEF HISTORY

When we look at a solid substance, such as a piece of iron, it appears to be continuous. This means that we might divide the piece of iron into smaller parts and then into still smaller parts, and repeat this process indefinitely and still have pieces of iron. No matter how small the piece of iron becomes, it would still be recognizable as a piece of iron. This concept, which might be called the concept of *continuous matter*, was widely held for a very long time. A number of Greek philosophers, including Aristotle, believed in this concept. Aristotle was so deeply respected that this concept, which is erroneous, was accepted by most people for hundreds of years.

However, there were a few philosophers among the ancient people who questioned this idea. One of these was the Greek philosopher Democritus, who thought that if matter were repeatedly subdivided, a particle would eventually be reached that could not be divided into smaller parts. As we know now, Democritus' idea was fundamentally sound; but he never gave experimental proof of his argument, and most people did not accept it.

3-2 THE ATOMIC THEORY

The *atomic theory* was stated by John Dalton shortly after 1800. He recognized that the generally accepted concepts of his time failed to account for many facts concerning matter which had been observed.

At that time a sufficient body of facts had been gathered to formulate two generalizations, or laws. One of these laws stated: "Matter can neither be created nor destroyed in a chemical reaction"; this is called the *law of conservation of matter*. As an example, if we start with known amounts of two different substances that will undergo chemical reaction and allow the reaction to take place, the total weight of the products formed will be exactly equal to the total weight of the materials that took part in the reaction, assuming, of course, that the reaction occurred in a closed system. This law had been carefully tested (to the best of their abilities) by the chemists of Dalton's time and was always found to be correct, within the limits of experimental error.

Another generalization had been made at Dalton's time, called the *law of definite composition*. This law states that any pure substance will always have exactly the same percentage composition. No matter what the source of pure water might be, it is always found on analysis to contain 88.8% oxygen and 11.2% hydrogen by weight. Similarly, any other pure substance will always have the same percentage composition, regardless of the source of the substance. This law was not consistent with the theory that matter was continuous. It seemed to indicate that there must be some small, indivisible particle in any element, and that the formation of a compound from two or more elements involved the combination of these tiny particles. At Dalton's time the behavior of gases had also been studied in some detail. These studies indicated that matter must be composed of small, discrete particles rather than being continuous.

From these two laws, Dalton conceived his theory, which stated the following three postulates:

1 Matter is not continuous but is composed of minute, indivisible particles, called *atoms*.

2 All of the atoms of an element are alike, and the atoms of one element are different from the atoms of another element. Since the atoms of an element

are alike, this also means that the atoms of an element would have the same weight, or at least the *same average weight*.

3   Compounds are formed by the combination of *whole* atoms of one element with a certain number of whole atoms of another element.

Using this theory, Dalton was able to account for the law of conservation of matter and also for the law of definite composition. Since Dalton's time, a great deal of additional evidence has been accumulated which indicates that his theory does indeed give a good picture of the *particulate nature of matter.*

One phenomenon that supports the concept of the particulate nature of matter is called *Brownian motion*. This motion can be observed if a suspension of very small smoke particles is observed under a microscope; these particles are observed to be in constant, irregular motion (see Fig. 3-1). It is easy to imagine that each smoke particle is being bombarded irregularly by still smaller, invisible particles, thus causing it to move about erratically. It is observed that the motion of the smoke particles becomes more vigorous and more erratic as the temperature increases. This is reasonable if we consider that a gas is composed of tiny, invisible particles that are in constant motion; these particles move faster as the temperature increases and therefore strike the smoke particles more frequently.

Another phenomenon which offers strong support to the theory of matter being composed of particles is observed with a cloud chamber. A simple cloud chamber is an enclosed space where the air is more than saturated with water vapor. If radioactive material is placed near a cloud chamber, "tracks" are ob-

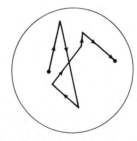

FIGURE 3-1 Possible pathway of smoke particle

served. These are believed to result from the condensation of tiny water droplets along the paths of particles given off by the radioactive material. Since this evidence indicates the radioactive material emits particles, it suggests that the material itself is composed of particles. There is little doubt today that matter is composed of submicroscopic particles since all the evidence from many independent sources point to the same conclusion.

The evidence for the particulate nature of matter tends to support Dalton's concept of the atom since the atoms envisioned by Dalton were extremely small particles. Most atoms have been found to have diameters that fall between 1 and 4 Å. (An angstrom unit, you will remember, is 1×10^{-10} m in length.) To point out the size of the atom in a different way, if you were dealing with atoms that were 1 Å in diameter, it would require about ten million of them laid side by side to reach across the head of an ordinary straight pin, which is roughly 1 mm across. The measurement of the diameter of an atom cannot be carried out by the use of ordinary methods of linear measurement, such as a foot ruler or yardstick. (The actual measurement of the diameter of the atom has been carried out in several different ways, all of which involve complex concepts that we shall not discuss.)

3-3 THE STRUCTURE OF THE ATOM

When Dalton's concept of the atomic nature of matter was finally accepted, it was believed that each of the atoms was like a hard little marble, with the atoms of each element like different kinds of marbles. This concept was generally accepted until the later part of the nineteenth century when a series of experiments led to the concept that atoms were composed of still smaller particles.

Experiments with evacuated tubes containing charged electrodes (gas-discharge tubes) were studied by Julius Plücker beginning about 1859. At this time positive and negative charges were known to exist. It was also known that like charges would repel each other and unlike charges would attract each other. (See Figure 3-2.) Plücker observed that in a partially evacuated tube containing high-voltage electrodes there was a stream of particles emitted at the cathode (*cathode rays*). (See Figure 3-3.) It was found that the cathode rays were identical when different gases were present in the tube. Identical particles were also emitted from different cathode materials. This led to the belief that identical subatomic particles were present in all types of atoms.

Cathode rays are now known to be streams of *electrons*. Further experiments by J. J. Thomson, R. A. Milliken, and others established the mass of an electron and its electric charge. Similar experiments with gas-discharge tubes using cathodes with holes drilled in them established the existence of positive particles. These were observed as rays traveling toward the cathode and through the holes (or canals) drilled in the cathodes. The mass of these particles differed with different gases in the tubes; and when hydrogen was used, the

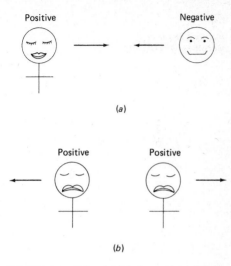

FIGURE 3-2 Behavior of charged particles: (a) attraction; (b) repulsion

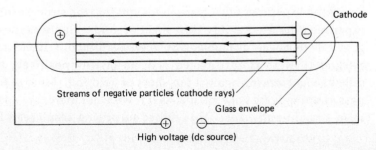

FIGURE 3-3 Cathode ray tube

TABLE 3-1 Subatomic particles

PARTICLE	SYMBOL	MASS, g	CHARGE	APPROXIMATE RELATIVE WEIGHT
Electron	e	9.11×10^{-28}	-1	$\frac{1}{1,800}$
Proton	p	1.67×10^{-24}	$+1$	1
Neutron	n	1.68×10^{-24}	0	1

proton was discovered. The *neutron*, an uncharged particle, was more difficult to pin down, but finally was detected by Chadwick in 1932.

These particles, the *proton*, the *electron*, and the *neutron* are now believed to be the fundamental particles present in the intact atom. The properties of these particles are noted in Table 3-1. Later work by nuclear physicists established the existence of a host of other subatomic particles; these are observed as products of the decomposition of atoms. The function of many of these particles is still being investigated.

The most important clue concerning the arrangement of these fundamental particles in the atom came from the work of Lord Rutherford. In 1911 he performed an experiment that gave an insight into the structure of the atom. In this experiment he used a very thin sheet of gold. This was the thinnest sheet of gold foil that was available, but it was still about 10,000 atoms or more in thickness. Rutherford directed a beam of tiny positively charged particles, called *alpha particles*, against the sheet of gold foil. Most of the alpha particles passed through the gold foil just as if it were not there; but a few were deflected from straight pathways. This rather surprising result lead Rutherford to the conclusion that atoms *were not* like hard little marbles, nor like a "raisin pudding," as others had supposed. He repeated the experiment many times and was able to determine with reasonable accuracy the number of alpha particles that passed straight through the gold foil and the number that were deflected from their pathways. In this way, he was able to show that there was a very tiny volume in the center of each atom in which almost all the weight of the atom was concentrated; this tiny, very dense portion of the atom Rutherford called the *nucleus*. Alpha particles were deflected from their course only when they approached this nucleus very closely. (See Figs. 3-4 and 3-5.)

Lord Rutherford was also able to show that the nucleus of the atom carried

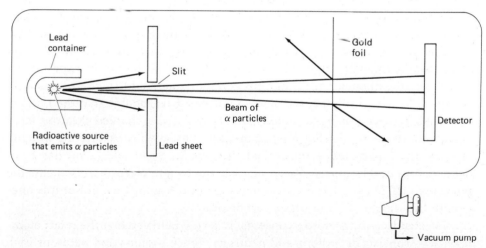

FIGURE 3-4 Rutherford's experiment

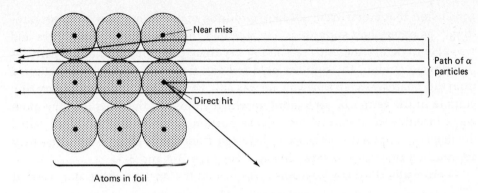

FIGURE 3-5 Atoms in gold foil

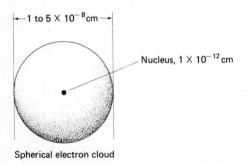

FIGURE 3-6 The nuclear atom

a positive charge of electricity because it would *repel* an alpha particle. Alpha particles were already known to be positively charged, and it was known that like charges of electricity would repel each other. Mathematical calculations show that the diameter of an atom is about 10^{-8} cm. Rutherford's experiments established the diameter of the nucleus at approximately 10^{-12} cm. Atoms were known to be electrically neutral, and so it seemed reasonable that the nucleus contained positively charged protons and the negatively charged electrons were outside the nucleus. (See Fig. 3-6.) In order to conveniently express the weights of these small particles, a unit called the *atomic weight unit* (awu) has been defined. This unit is approximately equal to the weight of a proton (actually, the proton is 1.00728 awu). The actual definition of the atomic weight unit is one-twelfth the weight of an ordinary carbon atom.

In extending Rutherford's concepts, it is now believed that the intact atomic nucleus consists of protons and neutrons. Since protons and neutrons each have a weight of about 1 awu, the weight of an atom is approximately equal to the total number of protons plus neutrons. The number of protons in the nucleus is called the *atomic number* of the atom. All atoms having the same number of protons are atoms of the same element. The atoms of a particular element may have different numbers of neutrons, however. Two atoms of the same element that have different numbers of neutrons are called *isotopes* of each other.

An atom will tend to interact with its surroundings through its outermost portions, or its electrons. Therefore the *chemistry* of an atom is largely determined by the structure of the *electrons* of the atom. A clue that led to a better understanding of the electron structure of the atom came from the study of the emission spectrum of hydrogen, which may be observed in Fig. 3-7.

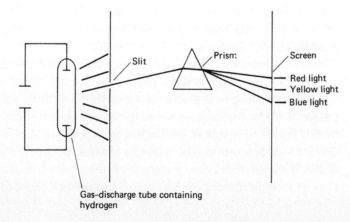

FIGURE 3-7 Emission spectrograph

Niels Bohr proposed a theory which quantitatively explained the hydrogen line spectrum. The basic ideas of Bohr's theory were as follows:

1 The electron is in orbital motion around the proton.

2 The energy of the electron is *quantized*, or restricted to certain values.

3 When an electron in a hydrogen atom changes from a higher energy state to a lower energy state, light is emitted. The emitted light has the same energy as the difference in energy of the two electronic energy states of the atom. (See Fig. 3-8.)

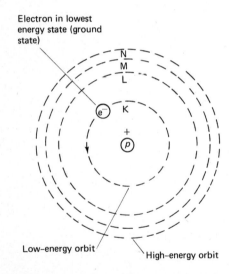

FIGURE 3-8 Bohr's concept of the hydrogen atom

Bohr's theory pictures the atom as being somewhat like the solar system, the planets being analogous to the electrons and the sun analogous to the nucleus of the atom. His theory also pictures more than one electronic energy state available to the electron. The lowest energy state is that in which the electron is closest to the nucleus; the higher energy states are those in which the electron is farther from the nucleus (Fig. 3-8).

When hydrogen is placed in a gas-discharge tube and an electric current is passed through the tube, some of the atoms become "excited"; that is, the electron is *promoted* to one of the higher energy states. Let us take, for example, an excited hydrogen atom which has its electron in the M shell rather than in the K shell. If the electron now moves from the M shell to the L shell (one possible change it could undergo), light which has energy equal to the energy difference in the two orbits ($E_M - E_L$) is emitted. It is known that this energy difference corresponds to light having a wavelength of 6,563 Å, or red light. (The energy of light is related to the wavelength by the following equation: $E = hc/\lambda$, where h is Planck's constant, c is the velocity of light, and λ is the wavelength of light.)

From his simple model Bohr was able to derive a mathematical equation which accurately predicted the observed lines in the spectrum of the hydrogen atom. Similar experiments with more complicated atoms suggested that some of Bohr's ideas could be extended to all atoms; presently accepted concepts of the structure of the different atoms grew out of Bohr's concepts.

3-4 THE ATOMIC STRUCTURE OF THE FIRST 20 ELEMENTS

Shortly after World War I, Irving Langmuir envisioned the atomic structure of the first 20 elements to be as shown in Fig. 3-9. (Diagrams of atoms that are used throughout this text are designed to be *useful learning tools* but do not represent the "true appearance" of the atoms; in fact, no one really knows what an atom looks like.) We observe that the hydrogen atom (element number 1) is the most simple atom, consisting of a single proton, which constitutes the nucleus of the atom, and an electron, which moves about outside of this nucleus. The symbol p is used to represent a proton, e is used to represent an electron, and n is used to represent a neutron.

The helium atom (element number 2) is the next most simple atom. In the nucleus of the helium atom there are two protons and two neutrons; outside of the nucleus there are two electrons. With these electrons the *first shell* of elec-

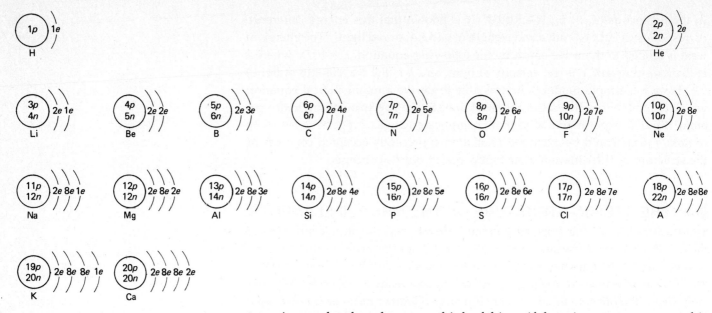

FIGURE 3-9 Atomic structure of the first 20 elements

trons is completed, and we may think of this as if there is no more room at this level for any more electrons.

Element number 3 is lithium. In the nucleus of the lithium atom there are three protons and four neutrons. The three electrons outside of the nucleus are arranged in such a manner that two of them are in the first shell of electrons (as in the helium atom) and the third electron is in a new shell, the second shell.

In Fig. 3-9, 2⟩e represents a shell containing two electrons, 3⟩e represents a shell containing three electrons, and so forth. We explain many observations by assuming that the *outer shell* (valence shell) of electrons never contains more than eight electrons; this is true of the heavier atoms also. Even those atoms with the largest number of electrons do not have more than eight electrons in the outer shell.

In studying Fig. 3-9, we observe that helium, neon, and argon have completed *outer shells of electrons*. This has been learned by experiment. In many experiments it has been observed that these elements will *not* undergo chemical reaction easily; that is, they are *chemically inert*. We may observe also that the other atoms that are shown do not have completed outer shells of electrons. Each of these atoms *will* undergo chemical reaction. And when these atoms undergo chemical reaction, they do so in order to achieve a *more stable arrangement of electrons*, that is, to reach an arrangement where *the outer shell of electrons is the same as an inert gas.*

At this point, we are prepared to make a general statement of great importance about chemical reactions. *Chemical reactions occur because of the tendency of atoms and molecules to go from less stable arrangements of electrons in their outer shells to more stable arrangements of electrons.* Other factors also influence the tendency of atoms and molecules to react, but this is the most important one.

3-5 ATOMIC WEIGHT AND ISOTOPES

The *atomic weight* of an element is the average weight of an atom of an element expressed in atomic weight units (awu). It was necessary to invent this atomic weight unit because of the difficulty of using other units for expressing the weight of the atoms and because the original determination of atomic weights was done by assessing the relative weights of the different atoms.

The atomic weight unit is defined as one-twelfth of the weight of one carbon 12 atom, for example, an atom containing six protons, six neutrons, and six electrons. This unit is equal to 1.66×10^{-24}g. The historical development of this unit, which involved some remarkable reasoning on the part of the early chemists, will not be discussed in this text but may be studied in texts dealing with the history of chemistry.

In discussing the atomic weights of the different elements, we have said that the atomic weight of an element is the *average* weight of an atom of that element. Let us consider more fully why the word *average* is needed in this statement. It has been found that in some cases the atoms of an element have differing numbers of neutrons; these atoms, identical in all respects except for the differing numbers of neutrons they contain, are called *isotopes*.

A sample of the element hydrogen actually consists of three different forms, or *isotopes*. The most abundant form of the hydrogen atom is the one that has only one proton in the nucleus of the atom (shown in Figs. 3-8 and 3-9). A small percentage of the hydrogen atoms that exist in nature also have one neutron in the nucleus in addition to one proton. Also relatively few hydrogen atoms exist that have two neutrons in the nucleus in addition to one proton (Fig. 3-10). These three different forms of hydrogen are called *isotopes of hydrogen*.

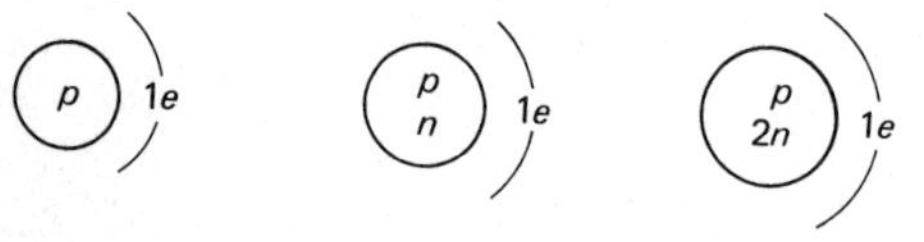

The atomic weight of hydrogen, 1.008 awu, is the average weight of the hydrogen atoms as they exist in nature. More than 99.9 percent of the hydrogen atoms have just one proton in the nucleus. In the case of the three hydrogen

FIGURE 3-10 The three isotopes of hydrogen

isotopes, the less common isotopes are given individual names. Thus, the isotope with one proton and one neutron in the nucleus is called *deuterium*, and the isotope with one proton and two neutrons is called *tritium*. This practice of giving individual names to the different isotopes of an element is not done with elements other than hydrogen. The exception in the case of hydrogen is due to the relatively large differences in properties caused by additional neutrons in the nucleus. The percentage change in weight is much greater in hydrogen than in other elements.

The most common method of designating the different isotopes is by means of *mass numbers* rather than by names. The mass number of an isotope is the total number of protons plus neutrons present in the nucleus of the atom. Thus, for the lightest hydrogen atom, the mass number is 1; for deuterium the mass number is 2; and for tritium the mass number is 3. The symbolism used is $_1^1H$, $_1^2H$, $_1^3H$; here the subscript is the atomic number and superscript is the mass number.

Let us consider another element which has more than one isotope: lithium. In one of the isotopes there are three protons and three neutrons in the nucleus. Here the mass number is 6, the isotope is called *lithium 6*, and the symbol is $_3^6Li$. Another isotope exists with three protons and four neutrons in the nucleus. This is called *lithium 7*, and the symbol is $_3^7Li$. The ratio of the natural occurrence of these isotopes is 8.3:91.7. In other words, 8.3 percent of the lithium atoms that occur in nature are lithium 6, ($_3^6Li$), and 91.7 percent are lithium 7 ($_3^7Li$). The average weight of the lithium atoms that occur in nature is 6.940 awu, and so this is the atomic weight of the element.

In the case of those elements where the atomic weight is appreciably dif-

TABLE 3-2 Comparison of sodium chloride with lithium chloride

PROPERTY	SODIUM CHLORIDE	LITHIUM CHLORIDE
Formula	NaCl	LiCl
Physical state	Solid	Solid
Color	White	White
Density	2.165 g/ml	2.068 g/ml
Boiling point	1413°C	1353°C
Melting point	801°C	613°C
Solubility	$\dfrac{35.7 \text{ g}}{100 \text{ ml}}$	$\dfrac{45.4 \text{ g}}{100 \text{ ml}}$

ferent from a whole number, there are relatively large proportions of different isotopes. The element chlorine, for example, which has an atomic weight of 35.457, is made up of 76 percent chlorine 35 and 24 percent chlorine 37.

3-6 ATOMIC STRUCTURE AND THE PERIODIC CHART

Referring again to Fig. 3-9, the atomic structure of the first 20 elements, we observe that a *periodic relationship* exists. This means that a certain aspect of the structure is *repeated periodically*. For example, we observe that the hydrogen, lithium, and sodium atoms each have one electron in the outer shell. Also the beryllium and magnesium atoms each have two electrons in their outer shells.

As we have mentioned, since the chemical behavior of an element depends on the outer shell of electrons, it is logical to expect that elements with the same numbers of electrons in their outer shells would exhibit similar chemical behavior. Since the outer shells of sodium and lithium are similar in structure, they react similarly and form similar compounds. Therefore, we would expect the compound lithium chloride to be much like sodium chloride (ordinary table salt). In fact, we can see their similarities in Table 3-2.

It was on the basis of this type of similarity that the periodic chart was originally devised. The first periodic charts were constructed without knowing the atomic structures of the elements but knowing only their properties. The first

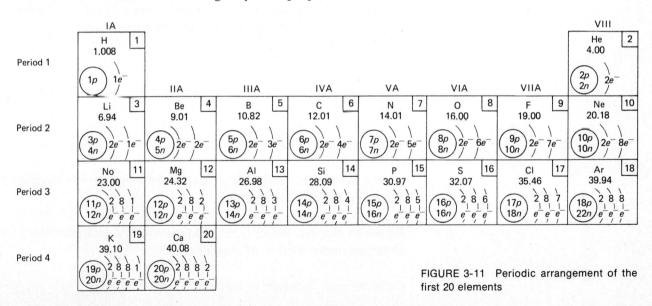

FIGURE 3-11 Periodic arrangement of the first 20 elements

periodic chart in approximately the modern form was developed by Mendeleev, a Russian chemist, in 1869. With our present knowledge of atomic structures, we can interpret the periodic properties in terms of *atomic structures*. The first 20 elements would be arranged in a periodic chart, as shown in Fig. 3-11. All the elements within a *group* have the same number of electrons in their outer shells and exhibit similar properties. The elements that are listed in a horizontal sequence constitute a *period*. All the elements within a *period* have the same number of shells of electrons.

As we continue our studies, we shall frequently refer to the material in this chapter on the structure of atoms and to the periodic chart. The material in this chapter may be said to be the "heart" of the study of chemistry.

GLOSSARY

Atom: The smallest particle of an element which is recognizable as that element

Atomic number: The number of protons in the nucleus of the atom

Atomic weight unit (awu): One-twelfth the weight of the most common type of carbon atom (see page 68)

Brownian motion: Random, zig-zag motion of a small particle which can be observed under a good microscope

Cathode rays: Streams of electrons

Conservation of matter: Matter can neither be created nor destroyed in a chemical change

Dalton's atomic theory: The first acceptable atomic theory of matter (see page 56)

Electron: A fundamental subatomic particle with charge of -1

Electron shell: The region outside of the nucleus in which electrons may be found

Group: The elements listed in a vertical column in the periodic chart (Fig. 3-11)

Isotopes: Atoms having the same atomic number but different mass numbers

Law of conservation of matter: There is no detectable weight change as a result of a chemical change.

Law of definite composition: A compound always has the same percentage composition

Mass number: The total number of protons plus neutrons present in the nucleus of the atom

Neutron: A fundamental subatomic particle with charge of zero (see page 60)

Nucleus: Contains the protons and neutrons of the atom and therefore most of the atom's mass (located in the center of the atom; see page 61)

Period: The elements listed in a horizontal row in the periodic chart

Proton: A fundamental subatomic particle with charge of +1, formed by taking an electron away from a hydrogen atom (usually found in the nucleus; see page 60)

Valence shell: The outer shell of electrons

SELF TEST

1. In a chemical reaction, the total weight of the products formed will be exactly equal to the total weight of the materials that took part in the reaction. This statement illustrates _____ .

2. Water is always 88.8% by weight oxygen and 11.2% by weight hydrogen. This statement illustrates _____ .

3. The random haphazard motion of small particles viewed under a microscope is called _____ .

4 Cathode rays are actually _____ .

5 The _____ of an element gives the number of protons in an atom of that element.

6 The _____ of an element gives the total number of neutrons plus protons in the nucleus of an atom of that element.

7 The outermost shell of electrons in an atom is called _____ and can contain no more than _____ electrons.

8 The isotope $^{38}_{18}$Ar contains _____ protons, _____ electrons, and _____ neutrons.

ANSWERS

1 The law of conservation of matter
2 The law of definite composition
3 Brownian motion
4 Streams of electrons
5 Atomic number
6 Mass number
7 The valence shell, eight
8 18, 18, 20

EXERCISES

1. Why were the ideas of Democritus never widely accepted?

2. State the law of definite composition.

3. State the law of the conservation of matter.

4. What are the three postulates of Dalton's atomic theory?

5. What observations support the idea that matter is composed of atoms?

6. What is the size range of atoms, expressed in (a) angstroms, (b) millimeters, (c) centimeters, and (d) inches? (*Hint:* Use the method of dimensional analysis for the conversion of units.)

7. Which of the subatomic particles is studied using a cathode ray tube?

8. Describe Rutherford's classical experiment which indicated that atoms were *not* solid bodies.

9. The theory proposed by Bohr pictured the atom as being analogous to what?

10. What three fundamental types of particles are now believed to make up all of the different types of atoms?

11. List the most important characteristics of the
    (a) Proton
    (b) Neutron
    (c) Electron

12. Will the individual electrons in the oxygen atom be identical with the electrons in a chlorine atom?

13. All atoms of the elements lithium, sodium, and potassium have (a) _____ electron(s) in their valence shell; all of these elements are in group (b) _____ of the periodic table. If a lithium atom lost an electron, the resulting particle would have the same arrangement of electrons as the inert element (c) _____ . This particle would have (d) _____ protons, (e) _____ neutrons, and (f) _____ electrons. Therefore the net electric charge of this particle would be (g) _____ .

14. Fluorine, chlorine, bromine, and iodine are all members of group (a) _____ . All atoms of these elements have (b) _____ electrons in their valence shell. If a chlorine atom gained an electron in the valence shell, the resulting particle would have (c) _____ electrons, (d) _____ protons, (e) _____ neutrons, and a net electric charge of (f) _____ . This particle would have the same arrangement of electrons as the inert element (g) _____ .

15 Beryllium, magnesium, and calcium are all members of group (a) _____ and have (b) _____ electron(s) in their valence shell. If a calcium atom lost two electrons, the resulting particle would have (c) _____ electrons, (d) _____ protons, and (e) _____ neutrons. The particle would have a net electric charge of (f) _____ and would have the same electron arrangement as the inert element (g) _____ .

16 Draw structures of the three isotopes of hydrogen.

17 Calculate the weight of one atom of sulfur (in grams).

18 Calculate the weight of 6.02×10^{23} iron atoms (in grams).

19 Calculate the weight of 6.02×10^{23} magnesium atoms (in grams).

20 Calculate the weight of the hydrogen atom (expressed in ounces).

21 Without referring to the text, draw the structures of the atoms of the first 20 elements.

22 How many electrons are contained in the outer shell of (a) oxygen, (b) aluminum, (c) carbon, and (d) argon?

23 What structural similarities exist between the elements within a *group* in the periodic chart (Figure 3-11)?

24 What structural similarities exist between the elements within a *period* in the periodic chart?

25 What does the mass number of an isotope represent?

26 Why was the atomic weight unit invented?

27 What does the atomic number of an element represent?

28 How does lithium 6 differ from lithium 7?

29 Calculate from the atomic number and the mass number the number of neutrons present in
    (a) Hydrogen 3
    (b) Lithium 7
    (c) Chlorine 35
    (d) Chlorine 37
    (e) Carbon 14

30 Write the symbols for the following isotopes using subscripts to in-

dicate the atomic numbers and superscripts to indicate the mass numbers:
(a) Chlorine 35
(b) Chlorine 37
(c) Carbon 12
(d) Carbon 14
(e) Uranium 238

31 How may the emission spectrum of hydrogen be related to the Bohr picture of the structure of an atom?

COMPOUNDS

COMPOUNDS

CHAPTER TOPICS
After mastering the material presented in this chapter, you should

Understand the difference between *elements* and *compounds*

Recognize that a compound does not show the same properties as its constituent elements

Understand what is meant by a *covalent bond*

Understand what is meant by *electrovalent bonding*

Know how to write *electron-dot structures* for simple compounds

Know how to predict the *electrovalence* of most of the first 20 elements

Understand the difference between an *atom* and an *ion*

Know the names, formulas, and electrovalences of seven very common *polyatomic ions*

4-1 COMMON COMPOUNDS

From a study of the elements and individual atoms, we next progress logically to a study of compounds. Although elements are the most simple substances, they are not the most commonly occurring. The substances that we encounter most frequently in our everyday lives are *compounds* or *mixtures of compounds*.

These compounds may be either *simple* or *complex*. Simple compounds are composed of only two or three elements that are combined in a manner which is easy to picture. Complex compounds may be composed of several elements that are combined in a manner which is more difficult to picture. (You will recall that a compound is defined as a substance that is composed of two or more elements chemically combined in a definite proportion by weight.) It must be emphasized that a compound is a *different substance* from any of the elements of which it is composed. In other words, the properties of a compound are different from the properties of any of the elements that have combined to form it.

An example of a simple compound that is very familiar is sodium chloride (table salt), which is composed of the elements sodium and chlorine. If we compare the properties of the individual elements sodium and chlorine with the properties of sodium chloride, we see the changes in properties that occur when this compound is formed from the elements.

Sodium is a soft silvery metal in the pure state; it melts at 97.5°C, boils at 880°C, and has a great tendency to undergo chemical reaction. When pure sodium is placed in contact with water, a violent reaction occurs, forming hydrogen gas and sodium hydroxide. Chlorine, on the other hand, is a greenish-yellow gas at room temperature. It can be condensed by cooling to a liquid that boils at −33.7°C and freezes at −102°C. Chlorine is a very reactive substance also and was used as a poisonous gas in World War I.

Sodium reacts with chlorine to form the compound sodium chloride. This is a compound that has properties that are markedly different from the properties of either sodium or chlorine. Sodium chloride is a white crystalline solid. It melts at 801°C and boils at 1413°C. It has little tendency to undergo chemical reaction; and instead of being poisonous (as are the individual elements), it is a necessary part of our diet. The differences between the properties of the elements and those of the compounds that they form are not always as spectacular as shown in this example, but there will always be definite differences.

Other simple compounds that we may frequently encounter are sodium bicarbonate, water, and carbon dioxide. From these examples we see that compounds may be solids, liquids, or gases at room temperature. Compounds may be necessary in the diet or poisonous to eat. It is impossible to tell whether a substance is a compound or an element by mere observation. The only way to determine that a substance is a compound is by actually separating it into the elements of which it is composed. The only way to determine that a substance is an element is by trying every known method of separating it into simpler substances. If no method is successful, we conclude that the substance is an element.

The separation of a compound into its constituent elements may be carried out in several ways, but never by *physical processes*, such as boiling, freezing, dissolving in a solvent, passing through a filter paper, or any mechanical means of separation. The separation of a compound into its elements can sometimes be done by heating the compound to a high temperature, causing *thermal decomposition*. In other cases a compound can be decomposed by passing a direct current of electricity through the material when it is melted. In still other cases a compound may be decomposed by means of chemical reactions involving other substances.

4-2 PARTICLES INVOLVED IN COMPOUND FORMATION

The formation of compounds from elements can be readily understood in terms of Dalton's theory of the atom. As an example, the element hydrogen combines with the element chlorine to form the compound hydrogen chloride. Gaseous hydrogen chloride has the following composition:

	% by weight
Hydrogen	2.74
Chlorine	97.26

Dalton's atomic theory can account for this composition if one hydrogen atom, which weighs approximately 1 awu, combines with one chlorine atom, which weighs approximately 35.5 awu, to give one *molecule* of hydrogen chloride (weighing 36.5 awu). (See Fig. 4-1.)

The percent by weight of hydrogen would be

$$\frac{1.0 \text{ awu}}{36.5 \text{ awu}} \times 100 = 2.74\%$$

Similarly, the percent by weight of chlorine is

$$\frac{35.5 \text{ awu}}{36.5 \text{ awu}} \times 100 = 97.3\%$$

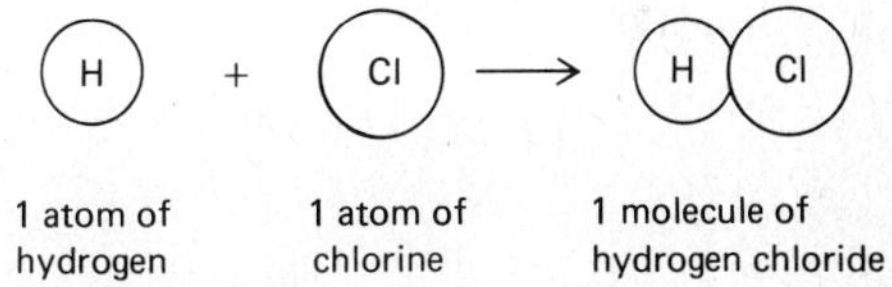

FIGURE 4-1 HCl formed from hydrogen and chlorine

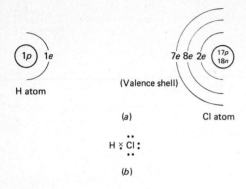

FIGURE 4-2 (a) Electronic configuration of hydrogen and chlorine; (b) electron-dot structure of hydrogen chloride

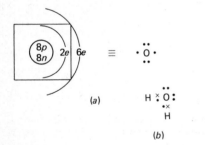

FIGURE 4-3 (a) Electron-dot representation of oxygen; (b) electron-dot representation of water

We have introduced the term *molecule* to describe the simplest particle of hydrogen chloride. A molecule is the simplest particle of a substance which is recognizable as that substance. It is the smallest particle of a compound that can exist.

4-3 COVALENT BONDING

The structure of both the hydrogen and chlorine atoms is shown in Fig. 4-2a. Note particularly the number of electrons in the outer shell (*valence shell*) of each atom. We can explain the combination by considering that they *share* a pair of electrons. This gives each atom the effective arrangement of electrons found in the nearest inert gas.

An alternative structure is shown in Fig. 4-2b. Here the dots represent the *valence electrons* (i.e., the electrons in the outer shell) of the chlorine atom, and the x represents the one valence electron of hydrogen. The shared pair of electrons constitutes what is known as a *covalent bond*. In covalent bonding, the electrons are shared by the two atoms. That is, the shared electrons are found in the vicinity of the chlorine atom part of the time and in the vicinity of the hydrogen atom part of the time. As a result, each atom achieves a stable arrangement of electrons. The chlorine atom effectively has the same arrangement of electrons as argon, and the hydrogen atom effectively has the same arrangement of electrons as helium.

Figure 4-2b, showing the valence electrons of the hydrogen and chlorine atoms, is called an *electron-dot structure*. In this structure the symbol of the element represents the nucleus of the atom and the inner stable shells of electrons. The electrons in the outer shell of each atom are represented by dots. Since electrons tend to occur in molecules in pairs, the dots are shown in pairs. Since electrons also tend to occur in octets, they are also shown in octets. To amplify further, we represent the structure of an oxygen atom in two ways, as shown in Fig. 4-3a. The water molecule (H_2O) then would be as shown in Fig. 4-3b.

A concept we have just introduced is that of *valence*. The valence of an atom was originally conceived as the combining capacity of that atom. For example, in the compound HCl each atom has combined with one other atom; therefore each atom has a valence of 1. In a water molecule, one oxygen atom has combined with two hydrogen atoms. The oxygen atom is exhibiting a valence of 2, and the hydrogen atom is exhibiting a valence of 1. Since the combination of atoms in these examples involves the sharing of electrons, the valence exhibited is called *covalence*.

Another example of a molecule in which atoms share electrons to form *covalent bonds* is carbon dioxide (CO_2). In a molecule of this compound, one carbon atom is combined with two oxygen atoms. The carbon atom has four valence electrons, and each oxygen atom has six valence electrons. In order that each atom achieve a stable arrangement of electrons, it is necessary to share four electrons between the carbon atom and each of the oxygen atoms (see Fig. 4-4). When four electrons are shared between two atoms, a *double bond* is produced. Thus the bonding in CO_2 would be represented as in Fig. 4-5. Each line between adjacent atoms represents a shared pair of electrons.

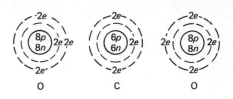

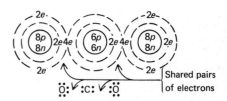

FIGURE 4-4 Electron distribution in carbon, oxygen, and carbon dioxide

FIGURE 4-5 Bonding in carbon dioxide

4-4 IONIC BONDING

Elements may combine to form compounds by a *transfer* of electrons. This produces a different type of compound, called an *ionic compound*. In this process the atom having a tendency to lose electrons (which usually is a metal) loses its valence electrons to the atom having a tendency to gain electrons (which

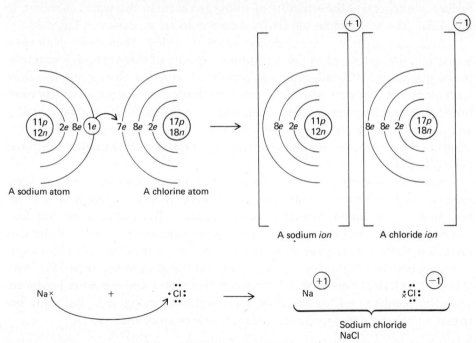

FIGURE 4-6 Electron distribution in sodium, chlorine, and sodium chloride

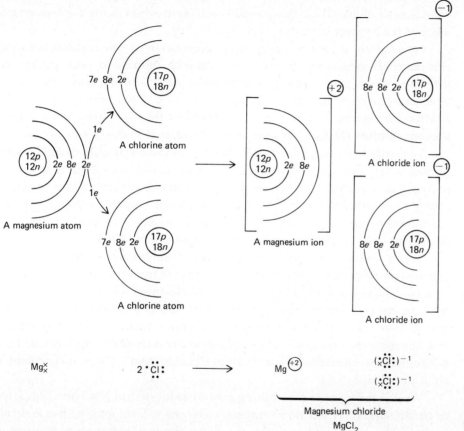

FIGURE 4-7 Electron distribution in magnesium, chlorine, and magnesium chloride

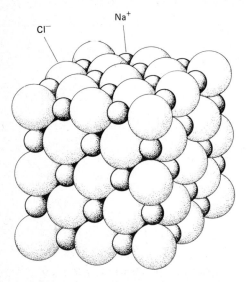

FIGURE 4-8 The arrangement of ions in NaCl

usually is a nonmetal). In this process each atom achieves the electron arrangement of the nearest inert gas.

For example, a sodium atom has one electron in its valence shell and a chlorine atom has seven. When sodium reacts with chlorine to form sodium chloride, the sodium atom loses an electron and becomes a sodium *ion*. The chlorine atom gains an electron and becomes a chloride *ion*. Each atom has achieved the electron arrangement of an inert gas. Together they constitute the compound sodium chloride (NaCl). This process is illustrated in Fig. 4-6.

To clarify further, an *ion* may be defined as the electrically charged particle formed when an atom (or group of atoms) gains or loses electrons to reach an inert-gas configuration. When an atom gains an electron, it is adding a negative charge to a neutral atom and so the resulting ion is negatively charged. When a neutral atom loses an electron, the removal of a negative charge leaves an excess of protons over electrons, forming a positively charged ion.

Let us consider the formation of another ionic compound from magnesium (a metal) and chlorine (a nonmetal). As you recall from the periodic chart (Table 2-4), the magnesium atom has two valence electrons which it tends to lose to form an inert-gas arrangement (neon). Each chlorine atom will gain one electron to form the electron arrangement of argon; therefore it would be expected that one magnesium ion would form for every two chloride ions formed. Figure 4-7 depicts the formation of magnesium chloride ($MgCl_2$) from magnesium and chlorine atoms.

We call the valence (combining power) exhibited in NaCl and $MgCl_2$ *ionic*, or *electrovalence*. When atoms transfer electrons to form ions, *no true molecule is formed*. The compound consists of a three-dimensional array of positive and negative ions stacked almost like oranges in a crate. Each positive ion is surrounded by negative ions, and each negative ion is surrounded by positive ions, as shown in Fig. 4-8. These ionic compounds conduct an electric current when they are dissolved in water solution because the ions are independently mobile. The positive ions migrate toward negative electrodes, and the negative ions migrate toward positive electrodes. A molten salt (a salt that has been melted) is also an excellent conductor of an electric current for the same reason.

In Fig. 4-9 we see a few more common electrovalences. The electrovalence of an element is the *charge of the ion* formed from an atom of the element. Note that the arrangement of elements in the periodic chart (Fig. 3-11) serves as an aid in remembering the structure of an atom and its electrovalence. The blank boxes indicate that the element belonging there seldom forms ionic bonds.

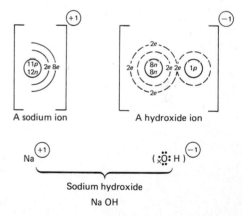

IA							
H^+	IIA	IIIA	IVA	VA	VIA	VIIA	VIIIA
Li^+	Be^{++}			N^{3-}	O^{--}	F^-	
Na^+	Mg^{++}	Al^{3+}		P^{3-}	S^{--}	Cl^-	
K^+	Ca^{++}						

FIGURE 4-9 Common electrovalences for the first 20 elements

4-5 COMPOUNDS WITH BOTH ELECTROVALENCE AND COVALENCE

Some compounds exhibit both electrovalence and covalence within the same compound. The covalent portion usually remains together as a unit, making it a *polyatomic ion* (an ion formed from more than one atom); and the unit carries a charge. For example, in the compound sodium hydroxide NaOH, there is a covalent bond between the oxygen and hydrogen atoms, making what is called a *hydroxide ion*. This group of atoms has gained one electron from the sodium. Thus the hydroxide ion carries an overall charge of -1, and the sodium ion carries a charge of $+1$. The attractive force between the sodium ion and the hydroxide ion is an *ionic bond* (Fig. 4-10).

FIGURE 4-10 Sodium hydroxide

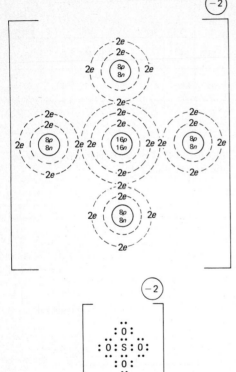

FIGURE 4-11 The sulfate ion and electron-dot structure

TABLE 4-1 Common polyatomic ions

FORMULA	NAME
NH_4^+	Ammonium
OH^-	Hydroxide
SO_4^{--}	Sulfate
NO_3^-	Nitrate
PO_4^{3-}	Phosphate
CO_3^{--}	Carbonate
$C_2H_3O_2^-$	Acetate

Another example of a *polyatomic ion* is the sulfate ion. This is formed from one sulfur atom, four oxygen atoms, and two electrons supplied by some metallic element. The sulfur atom shares electrons with the four oxygen atoms, and the two additional electrons give the ion a charge of -2 (see Fig. 4-11). You will notice that in Fig. 4-11 there are 50 electrons. Each oxygen atom has 8 electrons, and the sulfur atom has 16 electrons. Therefore, 48 electrons ($4 \times 8 + 1 \times 16$) come from the four oxygen atoms and the sulfur atom. (The two additional electrons have come from some metallic element, such as sodium.) The sulfate ion contains 48 protons and 50 electrons; therefore, it has a charge of -2.

As an example of a polyatomic ion that carries a positive charge, we shall consider the ammonium ion NH_4^+. The nitrogen atom shares one pair of electrons with each of the four hydrogen atoms. Since the nitrogen atom has five valence electrons by itself, there would be nine valence electrons available. But one electron is lost to some nonmetal, which leaves the ammonium ion with a total of one more proton than there are electrons, or charge of $+1$ (see Fig. 4-12).

Table 4-1 lists the most common polyatomic ions. These names, formulas, and electrovalences are used so frequently that they should be memorized. Throughout most of this text, polyatomic ions are written with parentheses

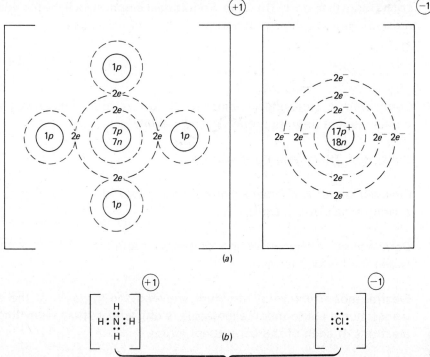

FIGURE 4-12 (a) Electronic structure of ammonium chloride (NH₄Cl); (b) electron-dot diagram of ammonium chloride.

around them for the sake of clarity and to emphasize the fact that this group of atoms often behaves as a unit. Once these polyatomic ions become thoroughly familiar, the parentheses are only needed in a formula where the polyatomic ion occurs more than once. (In Chap. 5 additional emphasis will be placed upon the nomenclature of compounds and the writing of formulas.)

GLOSSARY

Compound: A substance composed of two or more elements chemically combined in a definite proportion by weight

Covalence: The sharing of electrons

Covalent bond: A chemical bond between two atoms which is formed by sharing a pair of electrons

Double bond: A chemical bond formed by sharing two pairs of electrons between the bonded atoms

Electron-dot structure: A structure showing the symbols of the elements whose atoms are found in a molecule, along with dots representing valence electrons of each of the constituent atoms

Electrovalence: The charge of the ion formed from an atom of an element

Ion: The electrically charged particle formed when an atom or group of atoms gains or loses electrons to reach an inert-gas configuration

Ionic compound: A compound composed of ions

Molecule: The smallest particle of a compound that can exist

Percent by weight: The weight of the component divided by the weight of the whole (the resulting fraction is multiplied by 100)

Polyatomic ion: An ion formed from more than one atom

Valence: The combining capacity of an atom

Valence electrons: The electrons in the outer shell of an atom

SELF TEST

1. The bonds in compounds formed from a nonmetal combined with a metal are called (a) _____ bonds; the bonds formed between two nonmetals are generally (b) _____ bonds. A structure showing the arrangement of valence electrons in a molecule is called (c) _____.

2   Referring to the periodic chart, Table 2-4, write down the expected electrovalence of
    (a) Oxygen
    (b) Chlorine
    (c) Nitrogen
    (d) Magnesium
    (e) Sodium

3   Without referring to the text, write down the *correct* formulas *and* the electrovalences of the following polyatomic ions:
    (a) Carbonate
    (b) Hydroxide
    (c) Nitrate
    (d) Ammonium
    (e) Phosphate

4   Draw the electron-dot structures for the polyatomic ions listed in Prob. 3.

ANSWERS 1 (a) Ionic (or electrovalent)
            (b) Covalent
            (c) An electron-dot structure

2. (a) −2
   (b) −1
   (c) −3
   (d) +2
   (e) +1

3. (a) $(CO_3)^{--}$
   (b) $(OH)^-$
   (c) $(NO_3)^-$
   (d) $(NH_4)^+$
   (e) $(PO_4)^{3-}$

4. (a) 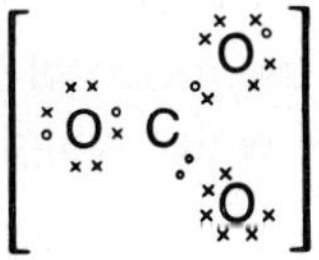 $\left[{}^{xx}_{xx}\text{O}{:}\text{C}{\genfrac{}{}{0pt}{}{{}^{x}_{x}\text{O}{}^{x}_{x}}{{}^{x}_{x}\text{O}{}^{x}_{x}}} \right]$

   (b) $\left[{}^{xx}_{xx}\text{O}{:}\text{H} \right]$

   (c) $\left[{}^{xx}_{xx}\text{O}{:}\text{N}{\genfrac{}{}{0pt}{}{{}^{x}_{x}\text{O}{}^{x}_{x}}{{}^{x}_{x}\text{O}{}^{x}_{x}}} \right]$

   (d) $\left[\begin{array}{c} \text{H} \\ \text{H}{:}\text{N}{:}\text{H} \\ \text{H} \end{array} \right]$

   (e) $\left[\begin{array}{c} {}^{xx}_{xx}\text{O}{}^{xx}_{} \\ {}^{xx}_{xx}\text{O}{:}\text{P}{:}\text{O}{}^{xx}_{xx} \\ {}^{xx}_{xx}\text{O}{}^{xx}_{} \end{array} \right]$

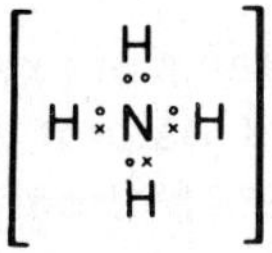

97
COMPOUNDS

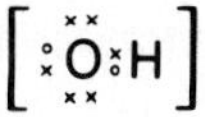

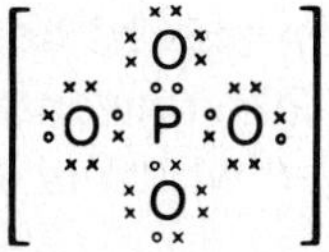

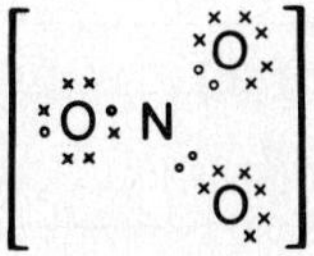

EXERCISES

1. How does a compound differ from an element?

2. Since we know that the element silver is a white shiny metal, is it possible for us to predict the properties of the compound silver nitrate?

3. Would it be possible to separate the chlorine from the sodium in sodium chloride by distillation?

4. A molecule of the compound carbon dioxide is formed from one atom of carbon and two atoms of oxygen.
   (a) Calculate the percent by weight of carbon in this compound.
   (b) Calculate the percent by weight of oxygen in this compound.

5. Calculate the number of grams of oxygen that can be obtained from the electrolysis of one glass of water if this amount of water weighs 250 g and the compound is 11.2% hydrogen.

6. How many pounds of iron may be obtained from 1 ton of iron ore if the ore is 80% hematite and 20% unwanted impurities? (Hematite is a compound of iron and oxygen which is 69.8% iron.)

7. Sodium is in group (a)_____, has (b)_____ valence electron(s), and usually shows an electrovalence of (c)_____.

8. Oxygen is in group (a)_____, has (b)_____ valence electron(s), and usually shows an electrovalence of (c)_____.

9. Chlorine is in group (a)_____, has (b)_____ valence electrons, and usually shows an electrovalence of (c)_____.

10. Magnesium is in group (a)_____, has (b)_____ valence electrons, and usually shows an electrovalence of (c)_____.

11. Sulfur is in group (a)_____, has (b)_____ valence electrons, and usually shows an electrovalence of (c)_____.

12. Bromine is in group (a)_____, has (b)_____ valence electrons, and usually shows an electrovalence of (c)_____.

13. Calcium is in group (a)_____, has (b)_____ valence electrons, and usually shows an electrovalence of (c)_____.

14. Fill in the formulas and electrovalences for the following polyatomic ions.
    (a) Hydroxide: (OH)‾
    (b) Carbonate: (   )‾
    (c) Acetate: (   )‾
    (d) Nitrate: (   )‾
    (e) Ammonium: (   )‾
    (f) Phosphate: (   )‾
    (g) Sulfate: (   )‾

15 Draw an electron-dot structure for the
(a) Chloride ion
(b) Hydroxide ion
(c) Lithium ion
(d) Calcium ion
(e) Neon atom
(f) Sulfide ion
(g) Chlorine molecule consisting of two chlorine atoms
(h) Phosphate ion
(i) Nitrate ion

16 Draw electron-dot diagrams of
(a) Water (H_2O)
(b) Carbon dioxide (CO_2)
(c) The fluoride ion

17 When a metal combines with a nonmetal, what type of bonding is formed?

18 Predict the type of bonding in each compound:
(a) NaF
(b) K_2S
(c) $MgCl_2$
(d) CO_2
(e) CCl_4

(f) Mg_3N_2
(g) NH_3
(h) SiO_2

19 Which of the substances in Exercise 18 should conduct an electric current when in the molten state?

20 Using the diagrammatic method for representing electronic structure (see Figs. 4-2a and 4-4), illustrate the electronic structure of the compounds in Exercise 18.

NOMENCLATURE AND CHEMICAL FORMULAS

CHAPTER TOPICS

There are two major goals to be achieved in studying this chapter. After reading it, you should be able to

Give a correct name to many fairly simple chemical compounds when confronted with the chemical formulas

Write a correct chemical formula (considering the valences of the elements) when given the name of a compound

In achieving these goals, you should note that there are two systems of nomenclature that will be considered: the older, traditional system; and a newer, systematic method of nomenclature, sometimes called the *Stock system*. The nomenclature used and the formulas presented in this chapter are examples of relatively simple *inorganic compounds*.

5-1 NOMENCLATURE OF A FEW SIMPLE COMPOUNDS

We have already encountered several formulas used to describe compounds. The complexity and composition of a chemical compound determines the way in which its name is derived. For simple *binary compounds* (compounds containing only two elements) the name of the compound is derived by naming the first element in the formula, then naming the second element in the formula, and then using the suffix *ide*. Let us first consider names of binary compounds in which a metal is combined with a nonmetal. For the compound NaCl, we

TABLE 5-1 Nomenclature of simple binary compounds

FORMULA	NAME
LiCl	Lithium chloride
KBr	Potassium bromide
$MgBr_2$	Magnesium bromide
CaO	Calcium oxide
Na_2O	Sodium oxide

TABLE 5-2 Nomenclature of binary compounds containing two nonmetals

FORMULA	NAME
CO	Carbon monoxide
SO_2	Sulfur dioxide
SO_3	Sulfur trioxide
NO_2	Nitrogen dioxide
SiO_2	Silicon dioxide

would use the name sodium chlor*ide*. Other examples of names of binary compounds are given in Table 5-1.

When both elements in a binary compound are nonmetals, we frequently use a *prefix* to indicate the number of atoms of the second nonmetal present. For example, the compound CO_2 is named carbon *di*oxide. Here the prefix *di* placed before the word *oxide* indicates the two oxygen atoms in the molecule. Other examples are given in Table 5-2. Compounds which contain polyatomic ions

TABLE 5-3 Nomenclature of compounds containing polyatomic ions

FORMULA	NAME
*$Na_2(SO_4)$	Sodium sulfate
*$K(NO_3)$	Potassium nitrate
*$Na_3(PO_4)$	Sodium phosphate
*$Ca(CO_3)$	Calcium carbonate
$Al(OH)_3$	Aluminum hydroxide
*$(NH_4)Br$	Ammon*ium* bromide
*$Na(C_2H_3O_2)$	Sodium acetate

*Parentheses are used in these formulas for emphasis only.

are named in the following way: First name the positive ion, and then name the negative ion. For example, Na(OH) would be named *sodium hydroxide*. Other examples are given in Table 5-3.

5-2 WRITING FORMULAS OF CHEMICAL COMPOUNDS

In writing correct chemical formulas for ionic compounds, one must remember that the total charge of a chemical formula must add up to *zero*. If we examine some of the examples in the preceding tables, we shall note that this rule has been followed. For example, in KBr the potassium ion carries a charge of +1, and the bromide ion carries a charge of −1; then +1 − 1 = 0. Sodium oxide is another example which illustrates the simple rule for writing chemical formulas of ionic compounds. The chemical formula of sodium oxide is Na_2O. Since the charge on the oxide ion is −2, it requires two sodium ions, each of which has a +1 charge. Summing the charges,

$$2 \times (+1) - 2 = 0$$

$$+2 - 2 = 0$$

In the case of compounds containing polyatomic ions, we follow the same rule, considering the polyatomic ion to be a simple charged particle. For example, in sodium sulfate, the sodium ion carries a charge of $+1$ and the sulfate ion carries a charge of -2. The formula is $Na_2(SO_4)$. In the case of calcium phosphate, the calcium ion carries a charge of $+2$ and the phosphate ion carries a charge of -3. In order to achieve an electrically neutral compound we need the same number of positive and negative charges. Thus we must use the least multiple of 2 and 3, or $2 \times 3 = 6$, as the total number of positive and negative charges. We can supply six positive charges by using three calcium ions, and we can supply six negative charges by using two phosphate ions. The formula of calcium phosphate, then, is $Ca_3(PO_4)_2$.

TABLE 5-4 Chemical formulas of some ionic compounds containing polyatomic ions

IONS IN COMPOUND	CORRECT FORMULA
K^+, $(CO_3)^{--}$	$K_2(CO_3)$
Mg^{++}, $(PO_4)^{3-}$	$Mg_3(PO_4)_2$
Ca^{++}, $(CO_3)^{--}$	$Ca(CO_3)$
$(NH_4)^+$, $(SO_4)^{--}$	$(NH_4)_2(SO_4)$
Na^+, $(NO_3)^-$	$Na(NO_3)$

TABLE 5-5 Some metallic ions that exhibit more than one electrovalence

ION	NAME
Cu^{++}	Cupr*ic*
Cu^{+}	Cupr*ous*
Sn^{4+}	Stann*ic*
Sn^{++}	Stann*ous*
Pb^{4+}	Plumb*ic*
Pb^{++}	Plumb*ous*
Hg^{++}	Mercur*ic*
*Hg^{+}	Mercur*ous*

*The mercurous ion actually exists as a diatomic ion Hg_2^{++} in compounds.

5-3 VARIABLE-VALENCE METALLIC IONS

Some metals are capable of existing in more than one electrovalence state. For example, iron is frequently observed to have an electrovalence of +2 or +3. In naming compounds containing iron it is necessary to distinguish between the two electrovalence states. The lower electrovalence of +2 is the ferr*ous* ion. The higher electrovalence of +3 is the ferr*ic* ion. A similar pattern is used in naming ions of other metals that have more than one electrovalence. (See Table 5-5.) The explanation of these observed electrovalences in terms of electronic structures is beyond the scope of this text. Additional common metallic ions that are frequently encountered in elementary chemistry courses are listed in Table 5-6.

A more systematic method of nomenclature has been devised, called the *Stock system*. In this system the charge on the metallic ion is indicated by a roman numeral in parentheses after the name of the metal. For example, $FeCl_2$

TABLE 5-6 Additional common metallic ions

ION	NAME
Zn^{++}	Zinc
Ba^{++}	Barium
Sr^{++}	Strontium
Ra^{++}	Radium
Cd^{++}	Cadmium
Ag^+	Silver
Al^{3+}	Aluminum

TABLE 5-7 Traditional and Stock-system names of a few typical compounds

FORMULA OF COMPOUND	TRADITIONAL NAME	STOCK NAME
$FeCl_2$	Ferrous chloride	Iron(II) chloride
$FeCl_3$	Ferric chloride	Iron(III) chloride
$Cu(NO_3)$	Cuprous nitrate	Copper(I) nitrate
$Cu(NO_3)_2$	Cupric nitrate	Copper(II) nitrate
SnF_2	Stannous fluoride	Tin(II) fluoride
SnF_4	Stannic fluoride	Tin(IV) fluoride
$Pb(SO_4)$	Plumbous sulfate	Lead(II) sulfate
$Pb(SO_4)_2$	Plumbic sulfate	Lead(IV) sulfate

is called *iron(II) chloride* because iron is in the $+2$ electrovalence state. The compound $FeCl_3$ would be called *iron(III) chloride*. Traditional and Stock-system names of a few typical compounds are given in Table 5-7.

The nomenclature which has been presented in this chapter does not apply to all possible compounds. For example, acids have a different nomenclature (discussed in Chap. 11). (Appendix VI lists additional kinds of nomenclature.)

5-4 FAMILIAR SUBSTANCES

Before a systematic way of naming compounds was developed, and even before chemical formulas of compounds were known, names were given to familiar substances. For example, water has been known for many years, but the chemical formula of water was not known until the nineteenth century. Table 5-8 lists some common names, chemical names, and chemical formulas of a few frequently encountered substances.

TABLE 5-8   Names of familiar substances

COMMON NAME	CHEMICAL NAME	CHEMICAL FORMULA
Water	Dihydrogen oxide	H_2O
Salt	Sodium chloride	$NaCl$
Milk of magnesia	Magnesium hydroxide	$Mg(OH)_2$
Alumina	Aluminum oxide	Al_2O_3
Litharge	Plumbous oxide	PbO
Slacked lime	Calcium hydroxide	$Ca(OH)_2$
Quick lime	Calcium oxide	CaO
Lye	Sodium hydroxide	$Na(OH)$

GLOSSARY

Binary compound: A compound containing two elements

Chemical nomenclature: A method of naming a chemical compound

Prefixes used in nomenclature:
Mono (one)
Di (two)
Tri (three)
Tetra (four)
Penta (five)

Stock system of nomenclature: A systematic method of nomenclature (see page 106-107)

Suffixes used in nomenclature:
ide Usually the ending of the name of a binary compound; also usually the ending of the name of a compound containing (OH).

ic Used to show the higher of two possible electrovalences of a metal (also has other uses: see nomenclature of acids, Chap. 11, Sec. 11-1)

ous Used to show the lower of two possible electrovalences of a metal (also has other uses: see nomenclature of acids, Chap. 11, Sec. 11-1)

PREPARATION FOR
GENERAL CHEMISTRY

SELF TEST 1 Write the correct formulas for the following compounds:
  (a) Sodium chloride   (i) Sodium hydroxide
  (b) Potassium bromide   (j) Sodium oxide
  (c) Magnesium oxide   (k) Ferric hydroxide
  (d) Iron(II) sulfate   (l) Cupric sulfate
  (e) Copper(II) hydroxide   (m) Stannous fluoride
  (f) Nitrogen trifluoride   (n) Mercuric chloride
  (g) Ammonium chloride   (o) Silver bromide
  (h) Nitrogen dioxide

2 Name the following compounds:
  (a) $HgCl_2$ (d) $CaCl_2$ (g) $Ca(OH)_2$ (j) $K_2(CO_3)$
  (b) $Fe(SO_4)$ (e) CO (h) $Fe_2(SO_4)_3$
  (c) Cu_2O (f) $Mg(NO_3)_2$ (i) $Hg_3(PO_4)_2$

ANSWERS 1 (a) NaCl   (e) $Cu(OH)_2$ (i) Na(OH)   (m) SnF_2
              (b) KBr   (f) NF_3 (j) Na_2O (n) $HgCl_2$
              (c) MgO   (g) $(NH_4)Cl$ (k) $Fe(OH)_3$ (o) AgBr
              (d) $Pb(SO_4)$ (h) NO_2 (l) $Cu(SO_4)$

2 (a) Mercury(II) chloride, or mercuric chloride
   (b) Lead(II) sulfate, or plumbous sulfate
   (c) Copper(I) oxide, or cuprous oxide

(d) Calcium chloride
(e) Carbon monoxide
(f) Magnesium nitrate
(g) Calcium hydroxide
(h) Iron(III) sulfate, or ferric sulfate
(i) Mercury(II) phosphate, or mercuric phosphate
(j) Potassium carbonate

EXERCISES

1 Complete the following table by writing in the formulas of compounds formed by the components indicated:

	$(NO_3)^-$	$(C_2H_3O_2)^-$	$(SO_4)^{--}$	$(CO_3)^{--}$	$(PO_4)^{3-}$
Na^+					
Mg^{++}		$Mg(C_2H_3O_2)_2$			
Al^{3+}					
Fe^{3+}					
Sn^{++}					
Pb^{4+}					

2. Name each of the compounds in Exercise 1 using the traditional system.

3. Name the compounds in Exercise 1 using the Stock system.

4. Name the following compounds:
   (a) KCl
   (b) NaBr
   (c) LiF
   (d) $BaCl_2$
   (e) $CaCl_2$
   (f) MgO
   (g) K_2O
   (h) HCl
   (i) HBr
   (j) Mg_3N_2
   (k) Na_2S
   (l) ZnS
   (m) $ZnCl_2$
   (n) Li_2O
   (o) $CaBr_2$
   (p) BaS
   (q) MgF_2
   (r) CaF_2
   (s) MgS
   (t) ZnF_2

5. Name the following compounds:
   (a) $Na_2(SO_4)$
   (b) Li(OH)
   (c) $Na(C_2H_3O_2)$
   (d) $K_2(CO_3)$
   (e) $(NH_4)_2(SO_4)$
   (f) $Ca_3(PO_4)_2$
   (g) $Mg(CO_3)$
   (h) $Ba(SO_4)$
   (i) $Be(NO_3)_2$
   (j) $Ca(C_2H_3O_2)_2$
   (k) $Li(NO_3)$
   (l) Na(OH)
   (m) $Ca(OH)_2$
   (n) $Ca(NO_3)_2$
   (o) $Zn(CO_3)$
   (p) $Al_2(SO_4)_3$
   (q) $Ba(NO_3)_2$
   (r) $Al(PO_4)$
   (s) $Mg(OH)_2$
   (t) $K_2(SO_4)$

6. Name the following compounds (either system is acceptable):
   (a) $HgCl_2$
   (b) Hg_2Br_2
   (c) PbO
   (d) PbO_2
   (e) $PbCl_2$
   (f) HgS
   (g) CuS
   (h) Cu_2S
   (i) $FeBr_2$
   (j) Fe_2O_3
   (k) $Pb(SO_4)$
   (l) $Pb(CO_3)$
   (m) $Hg(SO_4)$
   (n) $Hg_2(SO_4)$
   (o) $Cu(OH)_2$
   (p) $Fe_2(SO_4)_3$
   (q) $Sn(CO_3)$
   (r) $Fe(CO_3)$
   (s) $Sn_3(PO_4)_4$
   (t) $Sn(NO_3)_2$

7   Name the following compounds (note that each of these compounds contains two nonmetals):
    (a) CO_2 (d) SO_3 (g) P_2O_3*   (j) N_2O_4*
    (b) CO (e) N_2O*   (h) PCl_3 (k) NCl_3
    (c) SO_2 (f) P_2O_5* (i) NF_3

8   Draw electron-dot structures showing how the atoms share electrons for the compounds (a), (c), (d), (e), (h), (i), and (k) in Exercise 7.

9   Draw electron-dot structures showing the transfer of electrons for the following compounds:
    (a) $NaCl$ (c) $MgCl_2$ (e) AlF (g) Mg_3N_2
    (b) Na_2O (d) CaO (f) Li_2S (h) LiH

10 Write correct chemical formulas for the following compounds:
    (a) Potassium chloride   (i) Cupric oxide
    (b) Sodium sulfide   (j) Aluminum phosphate
    (c) Aluminum oxide   (k) Ammonium carbonate
    (d) Iron(II) sulfate   (l) Lead(II) acetate
    (e) Mercury(II) chloride   (m) Nickel(II) nitrate
    (f) Iron(II) oxide   (n) Calcium phosphate
    (g) Cuprous carbonate   (o) Cupric hydroxide
    (h) Copper(II) sulfate

*See Appendix VI for rules pertaining to these compounds.

11 Fill in the blanks:
   (a) FeCl$_3$ is _____ chloride.
   (b) Cu(SO$_4$) is _____ sulfate.
   (c) Na$_2$(CO$_3$) is sodium _____.
   (d) Sn(CO$_3$)$_2$ is _____ carbonate.
   (e) AlBr$_3$ is aluminum _____.
   (f) (NH$_4$)(NO$_3$) is _____ nitrate.
   (g) Ca(C$_2$H$_3$O$_2$)$_2$ is calcium _____.
   (h) Mg(OH)$_2$ is magnesium _____.

12 Fill in the blanks:
   (a) CO is carbon _____.
   (b) CO$_2$ is carbon _____.
   (c) N$_2$O$_4$* is dinitrogen _____.
   (d) P$_2$O$_5$* is diphosphorous _____.
   (e) N$_2$O$_3$* is dinitrogen _____.
   (f) P$_2$O$_3$* is diphosphorous _____.
   (g) SO$_3$ is sulfur _____.

13 Write the chemical formulas of these common substances:
   (a) Lye
   (b) Milk of magnesia
   (c) Alumina
   (d) Slacked lime
   (e) Salt

*See Appendix VI for rules pertaining to these compounds.

MATHEMATICS AND CHEMISTRY I

MATHEMATICS AND CHEMISTRY I

CHAPTER TOPICS

Quantitative concepts very crucial to understanding chemical calculations are presented in this chapter. You need to be able to do the following calculations and understand the following concepts:

Calculate the molecular weight from the correct chemical formula

Calculate the percent by weight of a given element in a compound from the correct chemical formula

Understand what a chemist means by the term *mole*

Calculate the number of grams in a mole of an element

Calculate the number of grams of an element in a given number of moles

Calculate the number of moles of an element in a given number of grams

Calculate the number of grams in a mole of a compound from the correct formula of the compound

Calculate the number of grams of a compound in a given number of moles of the compound

Calculate the number of moles of a compound in a given number of grams of the compound

Understand what is meant by the *empirical formula*

Understand what is meant by the *molecular formula*

Calculate the empirical formula from the percent composition of a compound

Calculate the molecular formula from the percent composition and molecular weight of a compound.

6-1 MATHEMATICAL CALCULATIONS AND THE STUDY OF CHEMISTRY

It is often necessary to apply mathematics in the study of chemistry. We have already encountered some of these mathematical calculations in dealing with the conversion of units in Chap. 1. As we progress in our studies, we shall find many additional opportunities to apply mathematics to chemistry. This means the chemistry student needs a solid foundation of fundamental mathematics. In the work that follows, some review of the principles of mathematical calculation is given. This will help the student understand more clearly the fundamental principles of mathematics that have been learned elsewhere. In addition, by the time the student completes this book, it is expected that he will be able to work many of the types of problems encountered in general chemistry with a minimum of difficulty.

We should recognize that our knowledge of a substance is always superficial and of a purely qualitative nature until we can make careful measurements of the quantities that are being investigated. The careful measurement of a particular quantity is frequently a time-consuming and expensive job. Usually such a measurement involves the use of equipment which may be quite costly.

If we go to the trouble of making a careful measurement of one quantity, we may wish to calculate the values of other quantities related to the first one. It is also possible, sometimes, to make calculations of quantities that are otherwise very difficult to measure. Thus we often have a need to apply mathematics to the study of chemistry.

6-2   CALCULATION OF MOLECULAR WEIGHT

One of the simple types of calculations used in chemistry is a calculation of the weight of one molecule of a compound. This is called the *molecular weight* of the compound. Let us review this calculation briefly: First, we recognize that a molecule of a compound is made up of atoms. Since this is true, we can easily calculate the weight of the molecule if we know the weights of the atoms. The weights of the atoms expressed in atomic weight units may be found written below the symbols of the atoms in the periodic chart (Table 2-4).

EXAMPLE   Calculate the weight of one molecule of water:

16.00 awu	(the weight of one oxygen atom)
1.01 awu	(the weight of one hydrogen atom)
1.01 awu	(the weight of one hydrogen atom)
18.02 awu	(the weight of one molecule of water)

We recall that the atomic weight unit is defined as one-twelfth of the weight of a $^{12}_{6}C$ atom.

EXAMPLE

The compound glycerin (also called glycerol) has the molecular formula $C_3H_8O_3$. Calculate the molecular weight of glycerin:

$3 \times 12.01 = 36.03$ awu	(the weight of three carbon atoms)
$8 \times 1.01 = 8.08$ awu	(the weight of eight hydrogen atoms)
$3 \times 16.00 = 48.00$ awu	(the weight of three oxygen atoms)
92.11 awu	(the weight of one glycerin molecule)

Therefore, the molecular weight of glycerin is 92.11 awu.

These two examples are of compounds containing only *covalent bonds*. In the case of an *ionic compound* a problem of terminology arises: In the development of the science of chemistry, there was a considerable period of time when all molecules were envisioned as what we would now call *covalent molecules*. With the advent of Arrhenius' theory of ionization during the 1880s, a problem with the term *molecular weight* arose. Since there is no true molecule of an ionic compound in the solid (or in solution), the term molecular weight is somewhat ambiguous. Sometimes the term *formula weight* is used instead to designate the total weight of the atoms represented by a chemical formula. For our present purposes, however, we shall use the term *molecular weight* to mean the total weight of the atoms in a chemical formula whether the compound is ionic or covalent.

EXAMPLE Calculate the molecular weight of aluminum sulfate $Al_2(SO_4)_3$:

$$2 \times 27.0 = 54.0 \text{ awu}$$
$$3 \times 32.1 = 96.3 \text{ awu}$$
$$3 \times 4 \times 16.0 = 192.0 \text{ awu}$$
$$\overline{342.3 \text{ awu}}$$

The molecular weight of $Al_2(SO_4)_3$ is 342.3 awu.

6-3 CALCULATION OF PERCENTAGE COMPOSITION

If the formula of a compound is known, it is not difficult to calculate the composition of the compound as percentages of the different elements *by weight*. Whenever we speak of percentages, we are referring to percentages by weight unless a definite statement of a different intent is made.

To calculate the percentage composition of a compound, first we calculate the molecular weight of the compound, or the weight of one molecule. Then we write down for each element a fraction that represents the fraction of the total weight of the molecule which is due to that element. Conversion of these fractions to percentages completes our calculations.

EXAMPLE Calculate the percentage of oxygen in a water molecule, H_2O:

Step 1 Calculate the molecular weight:

$$2 \times 1.01 = 2.02 \text{ awu}$$
$$1 \times 16.0 = 16.00 \text{ awu}$$
$$\overline{18.02 \text{ awu}}$$

Step 2 Calculate the weight fraction of the desired element:

$$\text{Weight fraction of oxygen} = \frac{1 \times 16.0 \text{ awu}}{18.02 \text{ awu}} = \frac{16.0}{18.0} = \frac{8}{9}$$

Step 3 Convert the weight fraction to a decimal fraction:

$$\frac{8}{9} = 0.888$$

Percent oxygen = $0.888 \times 100 = 88.8\%$

This means 88.8 percent of the weight of one water molecule is the weight of the oxygen atom. Since any amount of water is composed of many water molecules and each molecule is 88.8% oxygen, the entire quantity of water is 88.8% oxygen by weight.

We should bear in mind, however, that the element oxygen is present in water in a *combined form*. Although oxygen atoms are present in the water molecules, there are no free oxygen molecules or atoms. As we know, the compound water has properties that are very different from the properties of free oxygen.

EXAMPLE Calculate the percentage of nitrogen in calcium nitrate:

Step 1   Write the correct formula for calcium nitrate:

$$Ca(NO_3)_2$$

Step 2   Calculate the molecular weight:

$$\begin{array}{r} 1 \times 40.1 = 40.1 \text{ awu} \\ 2 \times 14.0 = 28.0 \text{ awu} \\ \underline{2 \times 3 \times 16.0 = 96.0 \text{ awu}} \\ 164.1 \text{ awu} \end{array}$$

Step 3   Calculate the weight fraction of the desired element:

$$\text{Weight fraction of } N = \frac{2 \times 14}{164.1} = \frac{28.0}{164.1}$$

Step 4   Convert the weight fraction to a decimal fraction:

$$\frac{28.0}{164.1} = 0.170$$

Percent nitrogen = $0.170 \times 100 = 17.0\%$

Now calculate percentages for each of the other elements in the previous example. As a check on your calculations, remember that the sum of the percentages should add up to approximately 100 percent.

6-4 THE MOLE CONCEPT

Since any weighable amount of a substance contains a huge number of atoms, molecules, or ions, it is convenient for the chemist to use a unit which represents a large number of particles. For example, it would be quite awkward for a chemist to talk about the number of dozens of atoms in 40.1 g of Ca because this amount of Ca contains 5.0×10^{22} doz. atoms. A chemist finds it most convenient to use the unit *mole:* One mole of atoms is 6.02×10^{23} atoms. Therefore, one could say that 40.1 g of Ca contains 1.0 mole of Ca atoms. (The number 1.0 mole is certainly less awkward to work with than 5.0×10^{22} doz.)

The number of particles in a mole, 6.02×10^{23}, is sometimes called *Avogadro's number,* which is defined as the number of $^{12}_{6}C$ atoms present in a sample containing exactly twelve grams of pure $^{12}_{6}C$. One mole of $^{12}_{6}C$ therefore weighs 12.0 g. The number of grams of an element required to contain one mole of atoms of that element can be determined easily.

EXAMPLE Calculate the number of grams of S which contains 1 mole of S atoms.

From the periodic table, the weight of one S atom = 32.1 awu. By definition, the weight of one $^{12}_{6}C$ atom = 12.0 awu. From the above, one can write

$$\frac{\text{Average weight of one S atom}}{\text{Weight of one } ^{12}_{6}C \text{ atom}} = \frac{32.1 \text{ awu}}{12.0 \text{ awu}}$$

Likewise

$$\frac{\text{Average weight of 1 doz S atoms}}{\text{Weight of 1 doz }^{12}_{6}\text{C atoms}} = \frac{12 \times 32.1 \text{ awu}}{12 \times 12.0 \text{ awu}} = \frac{32.1}{12.0}$$

Let us now consider 1 mole each of S and $^{12}_{6}$C:

$$\frac{\text{Weight of } 6.02 \times 10^{23} \text{ S atoms}}{\text{Weight of } 6.02 \times 10^{23} \text{ }^{12}_{6}\text{C atoms}} = \frac{6.02 \times 10^{23} \times 32.1 \text{ awu}}{6.02 \times 10^{23} \times 12.0 \text{ awu}}$$

$$= \frac{32.1}{12.0}$$

By definition, 1 mole of $^{12}_{6}$C weighs 12.0 g; therefore,

$$\frac{\text{Weight of 1 mole of S}}{12.0 \text{ g of }^{12}_{6}\text{C}} = \frac{32.1}{12.0}$$

Then the weight of 1 mole of S must be

$$\text{Weight of 1 mole S} = \left(\frac{32.1}{12.0}\right)(12.0 \text{ g}) = 32.1 \text{ g}$$

A very helpful shortcut in solving the above problem is easily realized: *The number of grams in 1 mole of an element is obtained by taking the numerical value of* the atomic weight of the element and assigning gram units to this number. For example, 1 mole of Fe would weigh 55.85 g, and 1 mole of Mg would weigh 24.3 g. This number of grams is sometimes called a *gram atomic weight* of the element, or, more simply, a *gram atom* of the element.

EXAMPLE

How many moles of Ca atoms are there in 23.0 g of Ca?

Step 1 Write the gram atomic weight of the element:

$$\text{Gram atomic weight} = 40.1 \text{ g}$$

Therefore, 1 mole of Ca = 40.1 g.

Step 2 Convert the given mass to moles:

$$23.0 \text{ g} \times \frac{1 \text{ mole}}{40.1 \text{ g}} = 0.573 \text{ mole}$$

Therefore, 23.0 g of Ca is 0.573 mole of Ca.

The following, more familiar, example will be solved in exactly the same way as the above example.

EXAMPLE

If 1 doz apricots (of uniform weight) weighs 40.0 oz, how many dozen apricots would there be in 10 oz of apricots?

PREPARATION FOR
GENERAL CHEMISTRY

$$1 \text{ doz} = 40 \text{ oz}$$

$$10 \cancel{\text{oz}} \times \frac{1 \text{ doz}}{40.0 \cancel{\text{oz}}} = 0.25 \text{ doz}$$

The following analogies can be made:

Mole ↔ dozen

Grams ↔ ounces

In chemical calculations, it sometimes is necessary to calculate the number of grams of an element from the number of moles, as illustrated in the following example.

EXAMPLE How many grams are there in 2.8 moles of Fe?

Step 1 Write the gram atomic weight of the element:

$$\text{Gram atomic weight} = 55.8 \text{ g}$$

Therefore, 1 mole of Fe = 55.8 g

Step 2 Convert moles into grams:

$$\left(2.8 \cancel{\text{moles}}\right)\left(\frac{55.8 \text{ g}}{1 \cancel{\text{mole}}}\right) = 156.0 \text{ g}$$

Therefore, 2.8 moles of Fe is 156.0 g.

The mole concept can be extended quite easily to problems involving molecules: One mole of molecules is 6.02×10^{23} molecules, and the number of grams in 1 mole of a compound can be easily determined from the chemical formula of the compound.

EXAMPLE

How many grams are there in 1 mole of water?

Step 1   Write the correct formula of the compound.

H_2O

Step 2   Figure out the molecular weight of the compound:

$2 \times 1.01 = 2.02$ awu
$1 \times 16.00 = 16.00$ awu
$\overline{18.02 \text{ awu}}$

Step 3   Assign gram units to the numerical value of the molecular weight. The result is called the *gram molecular weight* (the weight in grams of one mole of the compound). Therefore, 1 mole of water weighs 18.0 g.

EXAMPLE Calculate the number of moles of $C_{12}H_{22}O_{11}$ (table sugar) in 200.0 g of sugar:

Step 1 Calculate the gram molecular weight of sugar:

$$\begin{aligned} 12 \times 12.00 &= 144.0 \text{ awu} \\ 22 \times 1.01 &= 22.2 \text{ awu} \\ 11 \times 16.00 &= 176.0 \text{ awu} \\ \hline &\ 342.2 \text{ awu} \end{aligned}$$

Therefore, the gram molecular weight is 342; 1 mole = 342 g.

Step 2 Convert grams to moles:

$$\left(200.0 \text{ g}\right)\left(\frac{1 \text{ mole}}{342 \text{ g}}\right) = 0.585 \text{ mole}$$

Therefore, 200.0 g of $C_{12}H_{22}O_{11}$ contains 0.585 mole of sugar.

EXAMPLE How many molecules are there in 200.0 g of sugar?

Step 1 Realize that 1 mole contains 6.02×10^{23} molecules:

$$1 \text{ mole} = 6.02 \times 10^{23} \text{ molecules}$$

Step 2 Multiply the number of moles in the 200.0 g sample by the number of molecules in 1 mole:

$$(0.585 \text{ moles}) \left(\frac{6.02 \times 10^{23} \text{ molecules}}{1 \text{ mole}} \right) = 3.52 \times 10^{23} \text{ molecules}$$

Therefore there are 3.52×10^{23} molecules in 200.0 g of sugar.

The conversion of moles of a compound to grams can be done easily if the correct formula of the compound is known.

EXAMPLE

Alumina (Al_2O_3) is used in the manufacture of electrical insulators. How many grams of alumina are there in 0.68 mole of Al_2O_3?

Step 1 Find the molecular weight:

$$\begin{array}{r} 2 \times 27.0 = 54.0 \text{ awu} \\ 3 \times 16.0 = 48.0 \text{ awu} \\ \hline 102.0 \text{ awu} \end{array}$$

Therefore, 1 mole of Al_2O_3 = 102.0 g.

Step 2 Convert moles of the compound into grams:

$$(0.681 \text{ mole}) \left(\frac{102 \text{ g}}{1 \text{ mole}} \right) = 69.4 \text{ g}$$

Therefore, there are 69.4 g of alumina in 0.68 mole of Al_2O_3.

From these examples we see that the concept of a mole of material can be likened to the concept of a herd of cows. If we were to define a herd of cows as consisting of a certain definite number of cows (for example, 1,000 cows), it would be possible to *deal in units of herds*. Thus a farmer might be said to have 3 herds if he has 3,000 cows and $\frac{1}{10}$ of a herd if he has 100 cows.

As we progress with our study of chemistry we shall find that a thorough understanding of the term *mole* will be very useful. Many of the calculations that must be made can be carried out conveniently by employing this concept. In fact, it may be said that a person begins to think like a chemist when he begins to think in terms of moles.

6-5 EMPIRICAL AND MOLECULAR FORMULAS

The *empirical formula* of a compound gives the simplest ratio of atoms that are present in that compound. For example, the compound acetylene has one hydrogen atom for every carbon atom in the molecule. Thus the empirical formula is C_1H_1 (or just CH). It is known that a molecule of acetylene is composed of two carbon atoms and two hydrogen atoms. Therefore, the *molecular formula* of acetylene is C_2H_2. Note that the molecular formula is a multiple of the empirical formula. The *molecular formula* gives the actual number of atoms of each element necessary to form a molecule of a particular compound.

Empirical and molecular formulas do not appear by magic! For any compound they need to be determined by a chemical analysis of the substance followed by calculations of the formulas. An example of a chemical analysis for carbon and hydrogen would be as follows. Suppose that a sample of acetylene weighs 100 g. This represents the total weight of carbon plus hydrogen. The sample may be burned to yield carbon dioxide and water. Each carbon atom combines with oxygen from the air to form one molecule of CO_2, and every two hydrogen atoms combine with oxygen from the air to form one molecule of H_2O. The CO_2 formed can be collected and weighed, and then the weight of carbon that was present in the original acetylene may be calculated. The water formed can also be collected and weighed, and the weight of hydrogen present in the original acetylene sample can be calculated. From this, the ratio of carbon atoms to hydrogen atoms may be calculated.

EXAMPLE

A 100.0-g sample of acetylene is burned (reacts with oxygen) to yield 339 g of CO_2 and 69.3 g of H_2O. Calculate the empirical formula of acetylene.

First we must calculate the number of moles of the element carbon in 339 g of CO_2.

Step 1 Calculate the molecular weight of CO_2:

$$1 \times 12.0 = 12.0 \text{ awu}$$
$$2 \times 16.0 = \underline{32.0 \text{ awu}}$$
$$44.0 \text{ awu}$$

Step 2 Calculate the percent of carbon:

$$\text{Percent of C} = \frac{12.0 \text{ awu}}{44.0 \text{ awu}} \times 100$$

$$= 27.3\%$$

Step 3 Calculate the weight of carbon in 339 g of CO_2:

$$\text{Weight of C} = \left(\frac{27.3}{100.0}\right) \times 339 \text{ g} = 92.5 \text{ g}$$

Step 4 Calculate the number of moles of carbon in 92.5 g of carbon:

$$\text{Moles of C} = (92.5 \text{ g}) \left(\frac{1 \text{ mole}}{12.0 \text{ g}}\right)$$

$$= 7.71 \text{ moles}$$

Now we must calculate the number of moles of the element H in 69.3 g of H_2O.

Step 5 Calculate the molecular weight of H_2O:

$$\begin{array}{r} 2 \times 1.0 = 2.0 \text{ awu} \\ 1 \times 16.0 = 16.0 \text{ awu} \\ \hline 18.0 \text{ awu} \end{array}$$

Step 6 Calculate the percent of hydrogen in H_2O:

$$\text{Percent of H} = \frac{2 \times 1.0}{18.0} \times 100$$

$$= 11.1\%$$

Step 7 Calculate the weight of hydrogen in 69.3 g of H_2O:

Grams of H = (0.111) (69.3 g) = 7.70 g

Step 8 Calculate the moles of hydrogen in 7.70 g of H:

$$\text{Moles of H} = (7.70 \text{ g}) \left(\frac{1 \text{ mole}}{1.0 \text{ g}}\right) = 7.70 \text{ moles}$$

Now we must calculate the simplest whole-number ratio of C atoms to H atoms since this is defined as the empirical formula of the acetylene. The number ratio of C atoms to H atoms *is the same as* the ratio of moles of C to moles of H, or 7.71/7.70 ≅ 1:1. (The numbers 7.71 and 7.70 are close enough to the same value so that they can be considered equal.) The *empirical formula of acetylene* is therefore *CH*, and the *empirical formula weight* is 12.0 awu + 1.0 awu = 13.0 awu.

To determine the molecular formula of acetylene we must have one additional bit of information, namely, the *molecular weight*. The molecular weight of a gas may be determined experimentally. (The details of the calculation of molecular weight will be discussed in Chap. 13.) It has been determined that the molecular weight of acetylene is 26. To determine the molecular formula, we now divide the molecular weight by the empirical formula weight, for example, 26/13 = 2. This means that two empirical formulas are contained in one molecular formula. Therefore the *molecular formula* of acetylene must be $(CH)_2$ or C_2H_2.

We shall work through two more examples to illustrate the calculation of empirical and molecular formulas when the percent composition and molecular weight of the compound have already been calculated.

EXAMPLE A white crystalline solid is found by experiment to have a molecular weight of 180.0 awu and the following percent composition:

$$C = 40.0\%$$

$$H = 6.65\%$$

$$O = 53.3\%$$

Calculate the empirical and the molecular formulas of this compound.

Step 1 Assume that we have 100.0 g of the compound. In this amount we would have 40.0 g of combined C, 6.65 g of combined H, and 53.3 g of combined O.

Step 2 Calculate the number of moles of each element present:

$$\text{Moles of C} = (40.0 \text{ g}) \left(\frac{1 \text{ mole}}{12.0 \text{ g}}\right) = 3.33 \text{ moles}$$

$$\text{Moles of H} = (6.65 \text{ g}) \left(\frac{1 \text{ mole}}{1.0 \text{ g}}\right) = 6.65 \text{ moles}$$

$$\text{Moles of O} = (53.3 \text{ g}) \left(\frac{1 \text{ mole}}{16.0 \text{ g}}\right) = 3.33 \text{ moles}$$

Step 3 Write a *preliminary empirical formula*:

$$C_{3.33}H_{6.65}O_{3.33}$$

Step 4 From Dalton's theory, we know that only whole atoms can combine with one another. Therefore our empirical formula should contain whole-number subscripts. To convert the subscripts to whole numbers, we must divide by the smallest subscript in the preliminary empirical formula:

$$C_{3.33/3.33}H_{6.65/3.33}O_{3.33/3.33} = CH_2O$$

The *empirical formula* would therefore be CH_2O. The empirical formula weight would be

$$12.0 + 2.0 + 16.0 = 30.0 \text{ awu}$$

Step 5 Divide the molecular weight by the empirical formula weight:

$$\frac{180.0}{30.0} = 6.0$$

Therefore, six empirical formulas are contained in one molecular formula, and the *molecular formula* is $(CH_2O)_6$, or $C_6H_{12}O_6$.

The compound which we have described is commonly called *dextrose*, or *glucose*, the principal sugar found in human blood. When hospital patients are fed intravenously, they are usually given a 5% dextrose solution because dextrose provides a readily available source of energy.

Sometimes the ratio of atoms cannot be converted to a whole-number ratio in one step as in step 4 of the previous example. The following example illustrates such a problem.

EXAMPLE A gaseous hydrocarbon, found in "bottled gas," has a molecular weight of 44.0 awu and the following percent composition:

C = 81.8%

H = 18.2%

Calculate the empirical and molecular formulas.

Step 1 Assume that we have 100.0 g of the compound; in this amount we would have 81.8 g of combined C and 18.2 g of combined H.

Step 2 Calculate the number of moles of each element present:

$$\text{Moles of C} = (81.8 \text{ g}) \left(\frac{1 \text{ mole}}{12.0 \text{ g}}\right) = 6.82 \text{ moles}$$

$$\text{Moles of H} = (18.2 \text{ g}) \left(\frac{1 \text{ mole}}{1.0 \text{ g}}\right) = 18.2 \text{ moles}$$

Step 3 Write a preliminary formula:

$$C_{6.82}H_{18.2}$$

Step 4 Divide by the smallest subscript in the preliminary formula:
$$C_{6.82/6.82}H_{18.2/6.82} = CH_{2.66} \quad \text{or} \quad C_1H_{2\text{-}2/3}$$

Now we need to carry out an additional calculation in order to convert the subscripts to whole numbers, in accordance with Dalton's theory.

Step 4(a) Write the subscripts as fractions:

$$C_{1/1}H_{2\text{-}2/3} = C_{1/1}H_{8/3}$$

Rewrite the subscripts so that they have the same smallest denominators:

$$C_{3/3}H_{8/3}$$

Now multiply each subscript by the common denominator, which is 3 in this problem:

$$C_{3/3 \times 3}H_{8/3 \times 3} = C_3H_8$$

The *empirical formula* is therefore C_3H_8. The empirical formula weight is $3 \times 12.0 + 8 \times 1.0 = 44.0$ awu.

Step 5 Divide the molecular weight by the empirical formula weight:

$$\frac{44.0}{44.0} = 1.0$$

Therefore, one empirical formula is contained in a molecular formula. The *molecular formula* is just $(C_3H_8)_1$, or C_3H_8.

Remember the results of an empirical and molecular formula calculation can always be checked by making sure that the percent composition of the calculated formula agrees with the percent composition given in the problem. The molecular weight that is calculated for the molecular formula should agree, of course, with the given molecular weight.

GLOSSARY

Atomic weight unit: One-twelfth the weight of a $^{12}_{6}C$ atom = 1.67×10^{-24} g

Avogadro's number: The number of particles in a mole (equal to 6.023×10^{23})

Empirical formula: A formula that gives the simplest ratio of atoms that are present in a compound

Formula weight: The total weight of the atoms represented by a chemical formula

Gram atomic weight: The number of grams of an element which is composed of 6.02×10^{23} atoms of that element (numerically equal to the atomic weight)

Gram molecular weight: The weight in grams of one mole of a compound (numerically equal to the molecular weight)

Mole: A specific number (6.02×10^{23}) of particles

Molecular formula: A formula that gives the number of atoms of each element necessary to form one molecule of a compound

Molecular weight: The weight of one molecule of a compound expressed in atomic weight units

SELF TEST

1. Compute the molecular weight of magnesium sulfate.

2. What is the weight percent of phosphorus in Na_3PO_4?

3. What is the weight percent of nitrogen in $Ca(NO_3)_2$?

4. How many copper atoms are there in 0.50 mole of copper metal?

5. How many grams of copper are there in 0.50 mole of copper metal?

6   Compute the number of moles of magnesium in 57.0 g of magnesium metal.

7   The chemical formula of xylose, a type of sugar found in plants, has the chemical formula $C_5H_{10}O_5$. Compute the number of grams in 1 mole of xylose.

8   Compute the number of moles in 10.0 g of xylose.

9   Compute the number of grams of sulfur dioxide in 2.0 moles of SO_2.

10 Compute the number of moles of oxygen atoms required to form 2 moles of sulfuric acid (H_2SO_4).

11 What is the *empirical formula* of xylose ($C_5H_{10}O_5$)?

12 An oxide of nitrogen has the following percent composition: 30.5% N and 69.5% O. Compute the empirical formula of the compound.

13 If the oxide of nitrogen in Prob. 12 has a molecular weight of 92, compute the *molecular formula* of the compound.

ANSWERS

1. 120 awu
2. 18.9% phosphorous
3. 17.1% nitrogen
4. 3.01×10^{23} atoms
5. 31.8 g
6. 2.35 moles
7. 150 g
8. 0.0667 mole
9. 128 g
10. 8 moles
11. CH_2O
12. NO_2
13. N_2O_4

MATHEMATICS AND CHEMISTRY

EXERCISES

1. Why is the study of mathematics closely related to the study of chemistry?

2. If we do not make accurate measurements, we are learning only _____ information.

3. Define the atomic weight unit.

4. Define the term molecular weight.

5   Calculate the molecular weights of each of the following substances:
    (a) LiCl   (f) Al_2O_3
    (b) $CaBr_2$ (g) $SiCl_4$
    (c) CH_4O (h) $K(C_2H_3O_2)$
    (d) MgO   (i) $(NH_4)(NO_3)$
    (e) NH_4Cl (j) $Al_2(SO_4)_3$

6   Calculate the percentage composition by weight of each compound listed in Exercise 5.

7   Calculate the number of grams in
    (a) 1.2 moles of Fe
    (b) 0.53 moles of Pb
    (c) 6.2 moles of H_2O
    (d) 0.35 mole of $Na_2(SO_4)$
    (e) 12.5 moles of $C_6H_{12}O_6$

8   Hydrogen peroxide, used as a bleach and disinfectant, has the chemical formula H_2O_2.
    (a) What is its empirical formula?
    (b) What is its percentage composition?

9   A compound is found to have a molecular weight of 161.4 awu and to contain 40.4% zinc, 19.8% sulfur, and 39.6% oxygen.
    (a) How many grams of zinc exist in 1 mole of the compound?

(b) How many moles of zinc exist in 1 mole of the compound?
(c) How many moles of sulfur exist in 1 mole of the compound?
(d) How many moles of oxygen exist in 1 mole of the compound?
(e) Write the molecular formula of the compound.

10 Given the empirical formula and the molecular weight, calculate the molecular formula for the following empirical formulas:
(a) CH (molecular weight 78)
(b) CH_2O (molecular weight 180)
(c) CCl_3 (molecular weight 237)
(d) C_3H_2ClO (molecular weight 179)
(e) CF_2 (molecular weight 100)

11 In each of the following compounds, calculate the molecular formula of the substance:
(a) A compound with a molecular weight of 164.0, which contains 42.8% sodium, 18.9% phosphorus, and 39.0% oxygen.
(b) A compound with a molecular weight of 213.0, which contains 12.7% aluminum, 19.7% nitrogen, and 67.6% oxygen
(c) A compound with a molecular weight of 158.0, which contains 29.1% sodium, 40.5% sulfur, and 30.4% oxygen
(d) A compound with a molecular weight of 158.5, which contains 32.8% chromium and 67.2% chlorine
(e) A compound with a molecular weight of 60.6, which contains 20.0% carbon, 6.71% hydrogen, 46.65% nitrogen, and 26.64% oxygen

12 Ozone, whose chemical formula is O_3, is a particularly obnoxious component of polluted urban air. Compute the
   (a) Molecular weight of ozone
   (b) Number of moles of ozone in 25.0 g of O_3
   (c) Number of moles of oxygen atoms required to form 25.0 g of O_3
   (d) Number of grams of O_3 in 3.2 moles of ozone
   (e) Number of grams of O_3 that can be formed from 2.0 moles of O atoms

13 Fructose, an extremely sweet sugar, is used in preparing jam. It can be made by boiling fruits. The molecular weight of fructose is 180.0; its percent composition is 40.0% carbon, 6.67% hydrogen, and 53.3% oxygen. Calculate its molecular formula. (Note the similarity between glucose and fructose: Glucose and fructose have the same molecular formulas but they are *different* compounds. They are called *isomers*.)

14 In each of the following compounds, the percentage by weight of each element in the compound is given but the molecular weight has not been determined. Calculate the empirical formula of the compound, or the simplest ratio of atoms of each element in the compound if the compound is
   (a) 27.1% sodium, 16.5% nitrogen, and 56.4% oxygen
   (b) 29.2% nitrogen, 8.3% hydrogen, 12.5% carbon, and 50.0% oxygen
   (c) 36.8% iron, 21.2% sulfur, and 42.2% oxygen
   (d) 40.0% sulfur and 60.0% oxygen
   (e) 62.5% calcium and 37.4% carbon

15 Define the term mole.

16 How is 1 mole of a chemical compound similar to a herd of cows?

17 Calculate the number of moles of each compound if the given number of grams is present:
(a) 54.0 g of water
(b) 34.2 g of sugar (sucrose $C_{12}H_{22}O_{11}$)
(c) 80.0 g of calcium carbonate $Ca(CO_3)$
(d) 19.6 g of sulfuric acid $H_2(SO_4)$
(e) 196.0 g of phosphoric acid $H_3(PO_4)$
(f) 17.0 g of sodium nitrate $Na(NO_3)$
(g) 20.0 g of zinc sulfate $Zn(SO_4)$
(h) 50.0 g of sodium carbonate $Na_2(CO_3)$
(i) 35.0 g of lithium bromide
(j) 17.0 g of aluminum oxide
(k) 1.02 g of ferric hydroxide

18 Calculate the number of molecules of each compound listed in Exercise 17.

19 Grain alcohol has the chemical formula C_2H_5OH. Calculate the number of moles of each type of atom in 1.50 moles of pure C_2H_5OH.

20 The compound ammonium sulfate is often used as a nitrogen-containing fertilizer. Calculate the percent of nitrogen in ammonium sulfate

and compare it to the percent of nitrogen in ammonia (NH_3), which can also be used as a fertilizer.

21. Ascorbic acid (vitamin C) has a molecular weight of 176.1 and the following percent composition: 40.9% C, 4.58% H, and 54.5% O. Calculate its
   (a) Empirical formula
   (b) Molecular formula

22. An oxide of nitrogen, which is used as a component of rocket propellant, has a molecular weight of 92.0 and is 30.5% N and 69.5% O.
   (a) Determine the molecular formula of the compound.
   (b) Name the compound.

23. Write the empirical formulas for the following compounds:
   (a) C_6H_6
   (b) C_2H_6
   (c) P_4O_{10}
   (d) $C_5H_{10}O_5$
   (e) $C_6H_6Cl_6$
   (f) Hg_2Cl_2

CHEMICAL REACTIONS AND CHEMICAL EQUATIONS

CHEMICAL REACTIONS AND CHEMICAL EQUATIONS

CHAPTER TOPICS

As you may well have realized, learning chemistry has some similarity to learning a foreign language. New symbols and new words must be learned. Chemical symbols and chemical formulas form the basis of a *chemical language*. As you read through this chapter, you will learn

What a chemical reaction is

How to translate a word statement describing a chemical reaction into a *chemical equation*

Why a chemical equation must be balanced

How to balance a chemical equation

Some of the basic types of chemical reactions

7-1 CHEMICAL REACTIONS

In Chap. 4 we discussed chemical compounds and the formulas which the chemist uses as a shorthand to represent these compounds. If we think back to our original definition of chemistry, we remember that it is a study, not only of compounds, but also of changes that substances will undergo. These changes are called *chemical reactions*. We can say that whenever one type of substance undergoes a change and is converted into *another type* of substance, a chemical reaction has taken place.

A familiar chemical reaction is burning, or *combustion*. When natural gas [which is mostly methane (CH_4)] burns, a change occurs, and substances are formed which have properties that are different from those of methane. By chemical analysis these substances have been identified as carbon dioxide and water, neither of which will burn.

Another example of a chemical reaction is the *electrolysis* of water. If a direct current of electricity is passed through water (which contains a small amount of sulfuric acid), the water decomposes into two gases. Chemical analysis of the gases shows them to be hydrogen and oxygen. Hydrogen gas and oxygen gas are obviously different from the original water. This decomposition of water indicates that a chemical reaction has occurred. Electrolysis of water occurs when an automobile battery is "overcharged." The gases that are observed rising near the plates of the battery are hydrogen and oxygen. Since this mixture of gases is an explosive one, overcharging is a potentially hazardous situation. Other familiar examples of processes in which chemical reactions occur are the rusting of iron, the tarnishing of silver or copper, and the digestion of food.

In dealing with chemical reactions, the question often arises, "How do you know what will be formed as a result of this reaction?" Actually (and this may come as a shock to many people), the science of chemistry has not yet reached the stage where chemists can always predict whether or not a reaction will occur or what the results of a reaction will be. The only way to know the results is by experiment! In other words, someone must try the process.

As we proceed with our study of chemistry, however, we will find that we can often make intelligent predictions about chemical reactions if we have learned what has happened in similar types of reactions.

7-2 WRITING A CHEMICAL EQUATION

The chemical reactions which we have described may also be represented by a chemist's shorthand, which is called a *chemical equation*. In representing the

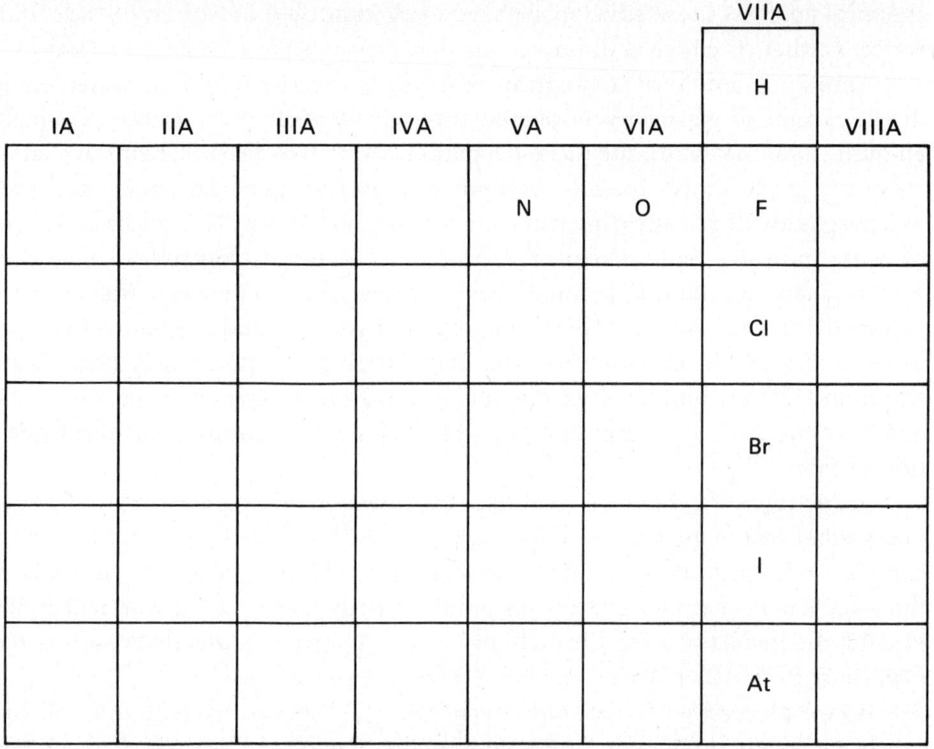

FIGURE 7-1 Elements that occur naturally with two atoms per molecule (diatomic elements)

chemical reaction of electrolysis of water, first we write the *correct* chemical formula of water (H_2O), which is the *reactant* in our example. Next we write an arrow →, which may be read as "yields." Then we write the correct formulas for the *products* of the reaction to the right of the arrow. If an element is formed which exists naturally with two atoms in each molecule, the symbol of the element is followed by the subscript 2 (see Fig. 7-1). Each chemical formula of the products (the substances formed in a chemical change) is separated by a plus sign (+). The *unbalanced chemical equation* is thus

$$H_2O \rightarrow H_2 + O_2$$

Since we know that atoms of elements are neither created nor destroyed in a chemical change, we must *balance* the chemical equation. This is done by writing the appropriate coefficients *in front of* the formulas. The coefficients must be chosen so that the same number of atoms of each element appears on each side of the arrow. Thus the correct *balanced chemical equation* for electrolysis of water is

$$2H_2O \rightarrow 2H_2 + O_2$$

To illustrate, let us use the symbol ⊗ to represent an atom of hydrogen and the symbol ● to represent an atom of oxygen. A diagrammatic representation of the process is shown in Fig. 7-2. From the figure it is apparent that the two oxygen atoms needed to form one molecule of the element cannot be obtained from a single water molecule. Thus, in balancing the equation, it is necessary to use

FIGURE 7-2 Unbalanced equation

two water molecules to supply the two oxygen atoms. The balanced equations are represented in Fig. 7-3.

As another example, let us write the chemical equation which describes the burning of methane. First we write the correct chemical formulas of the starting substances (reactants) and products:

$$CH_4 + O_2 \rightarrow CO_2 + H_2O \quad \text{(unbalanced equation)}$$

Now the equation must be balanced:

$$CH_4 + 2O_2 \rightarrow CO_2 + 2H_2O \quad \text{(balanced equation)}$$

7-3 TYPES OF CHEMICAL REACTIONS

Many (but not all) chemical reactions may be represented by a few simple types of reactions. Learning to recognize these types of reactions will help us predict the products that may be formed when a chemical change occurs. A list and brief description of five types of chemical reactions is given in Table 7-1.

TABLE 7-1 Types of chemical reactions

Combination	In the simple case, two elements combine to form a compound; for example, carbon reacts with oxygen to yield carbon dioxide, $C + O_2 \rightarrow CO_2$. In other cases, molecules may combine to form more complex molecules. For example, carbon dioxide reacts with water to form carbonic acid, $CO_2 + H_2O \rightarrow H_2CO_3$.

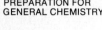

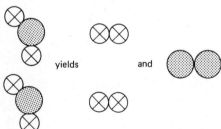

FIGURE 7-3 Balanced equation

Decomposition	A compound is decomposed into simpler substances. For example, water is decomposed into hydrogen and oxygen, $2H_2O \rightarrow 2H_2 + O_2$.
Displacement	One element displaces another element from a compound. For example, zinc reacts with hydrogen chloride to yield hydrogen and zinc chloride, $Zn + 2HCl \rightarrow H_2 + ZnCl_2$.
Exchange	The elements comprising one compound are exchanged with the elements of another compound. For example, sodium carbonate reacts with calcium chloride to yield calcium carbonate and sodium chloride: $Na_2(CO_3) + CaCl_2 \rightarrow Ca(CO_3) + 2NaCl$.
Combustion	The reaction of a substance with oxygen in the atmosphere, usually accompanied by release of heat and light. For example, acetylene reacts with oxygen to yield carbon dioxide and water, $2C_2H_2 + 5O_2 \rightarrow 4CO_2 + 2H_2O$. (*Note*: The *combustion* of an element can be correctly classified as a *combination*.)

7-4 EXAMPLES OF EQUATION WRITING

Many familiar processes can be considered to further illustrate the technique of writing *chemical equations*. When an aluminum can is exposed to air over a long period of time, a *combination reaction* with oxygen occurs. The process can be described as "Aluminum plus oxygen yields aluminum oxide." Translating this statement into the chemist's shorthand (correct formulas) gives

$Al + O_2 \rightarrow Al_2O_3$ (unbalanced equation)

At this point we must emphasize the importance of writing the chemical formulas *correctly* before attempting to balance an equation. Concentrating carefully on obtaining the correct formulas before proceeding further can make writing chemical equations a meaningful exercise. If the chemical formulas are not written *correctly*, the entire equation will become meaningless. (In order to write correct chemical formulas, the student must know the correct valences of atoms and ions. Valences were studied in Chap. 5, and if that material has not yet been mastered, a careful review of Chap. 5 is advisable at this time.)

Most chemists, after a little experience, are able to balance equations intuitively without actually thinking through each process. For the neophyte, however, the process can be very frustrating. Different types of equations actually require slightly different approaches.

We shall attempt to follow through the mental process for balancing the equation given above. First we observe that neither aluminum nor oxygen shows the same number of atoms on both sides of the equation. The ease of balancing an equation depends somewhat on which element is balanced first. However, if the wrong choice is made at first, it can be corrected later. In our example, the number of oxygen atoms should be balanced first since the formula of aluminum metal is just Al. The number of aluminum atoms can be easily adjusted since the formula shows just one atom per molecule. To balance the number of oxygen atoms, we observe that the total number of oxygen atoms on each side of the equation must be divisible by both 2 and 3 (that is, the least multiple of 2 and 3), or $2 \times 3 = 6$. To supply six oxygen atoms using O_2 molecules, $\frac{6}{2}$, or *three*, molecules of O_2 are required; and to supply six oxygen atoms using Al_2O_3, $\frac{6}{3}$, or *two*, molecules of Al_2O_3 are required. Inserting these coefficients gives us

$$Al + 3O_2 \rightarrow 2Al_2O_3 \quad \text{(unbalanced equation)}$$

Now we observe that 2×2, or 4, aluminum atoms are needed. Therefore, we write as the complete balanced equation

$$4Al + 3O_2 \rightarrow 2Al_2O_3 \quad \text{(balanced equation)}$$

In general, it will be easier to complete the balancing of an equation if that element which requires the determination of a least common multiple is balanced first.

Another familiar example of a combination reaction is the tarnishing of a copper-bottomed pot when it is heated. This involves the reaction of copper metal with oxygen gas in the atmosphere. The word equation can be written as

Copper metal plus oxygen gas yields copper (II) oxide

The *unbalanced* chemical equation is

$$Cu + O_2 \rightarrow CuO$$

and the *balanced* chemical equation is

$$2Cu + O_2 \rightarrow 2CuO$$

An example of a *decomposition reaction* occurs in the conversion of bauxite ore (Al_2O_3) into aluminum metal and oxygen gas. The word equation is

Aluminum oxide decomposes to yield aluminum metal plus oxygen gas

The *unbalanced* chemical equation is

$$Al_2O_3 \rightarrow Al + O_2$$

and the *balanced* chemical equation is

$$2Al_2O_3 \rightarrow 4Al + 3O_2$$

Another familiar example of decomposition occurs when hydrogen peroxide is applied to a cut. The word equation is

Hydrogen peroxide decomposes to yield water and oxygen gas

The *unbalanced* chemical equation is therefore

$$H_2O_2 \rightarrow H_2O + O_2$$

and the *balanced* chemical equation is

$$2H_2O_2 \rightarrow 2H_2O + O_2$$

If you have ever applied hydrogen peroxide to a cut, you have probably noticed that bubbles form. These bubbles are the oxygen gas that is generated from the decomposition reaction.

A method of recovering copper metal from the dissolved copper salts found in the water flow emerging from a copper mine is an example of a *displacement reaction*. The word equation is

Copper (II) sulfate plus iron yields copper plus iron (II) sulfate

and the chemical equation (balanced) is

$$Cu(SO_4) + Fe \rightarrow Cu + Fe(SO_4)$$

Writing the correct chemical formulas in this case provides a *balanced* chemical equation.

Another example of a displacement reaction is when precious metals are recovered from liquid solutions by displacement with zinc. The word equation is

Zinc metal reacts with silver nitrate to yield zinc nitrate plus silver metal

The *unbalanced* chemical equation is

$$Zn + Ag(NO_3) \rightarrow Zn(NO_3)_2 + Ag$$

and the *balanced* chemical equation is

$$Zn + 2Ag(NO_3) \rightarrow Zn(NO_3)_2 + 2Ag$$

Notice that the equation can be easily balanced by treating the nitrate ion as a unit. Polyatomic ions are usually handled this way in equation writing.

As an example of an *exchange reaction*, consider what happens when vinegar, which is a water solution of hydrogen acetate, is mixed with sodium bicarbonate. The word equation is

> Hydrogen acetate reacts with sodium bicarbonate to yield hydrogen carbonate plus sodium acetate

The chemical equation is

$$H(C_2H_3O_2) + NaH(CO_3) \rightarrow H_2(CO_3) + Na(C_2H_3O_2)$$

The equation is *balanced* as written. However, the hydrogen carbonate is an unstable compound which decomposes to yield water and carbon dioxide:

$$H_2(CO_3) \rightarrow H_2O + CO_2$$

Here the polyatomic ion decomposes.

Another example of an exchange reaction is the process in which milk of magnesia reacts with stomach acid (a water solution of hydrogen chloride). The word equation is

Magnesium hydroxide plus hydrogen chloride yields water plus magnesium chloride

and the balanced chemical equation is

$$Mg(OH)_2 + 2HCl \rightarrow 2H(OH) + MgCl_2$$

This example illustrates a reaction between an *acid* and a *base*. Hydrogen chloride dissolved in water is called *hydrochloric acid*. The most common acids contain the H^+ ion, and the most common bases contain the $(OH)^-$ ion. (These types of compounds are discussed further in Chap. 11.)

The last type of reaction that will be illustrated is *combustion*. A common type of combustion reaction occurs when a hydrocarbon (a compound containing only C and H) burns. A word equation describing such a process is

Butane reacts with oxygen in the air to yield carbon dioxide plus water

The unbalanced chemical equation is

$$C_4H_{10} + O_2 \rightarrow CO_2 + H_2O$$

and the balanced chemical equation is

$$2C_4H_{10} + 13O_2 \rightarrow 8CO_2 + 10H_2O$$

You will note that the first attempt to balance this equation would probably give

$$C_4H_{10} + \tfrac{13}{2} O_2 \rightarrow 4CO_2 + 5H_2O$$

In most cases, however, whole-number coefficients are more desirable. The entire equation is multiplied by 2 in order to remove the fraction $\tfrac{13}{2}$.

To summarize, in writing an equation for a chemical reaction:

1. Write the correct formulas of the reactants, considering the valences of the elements in doing so.
2. Write an arrow.
3. Write the correct formulas of the products.
4. Balance the equation.

When writing equations it is sometimes helpful to include some information about the physical conditions involved in the process. For example, the symbol Δ is used to indicate that the reactants must be heated to cause the reaction to occur. Such an equation might be written

$$2HgO \xrightarrow{\Delta} 2Hg + O_2$$

The word equation would be:

Two molecules of mercury(II) oxide are heated and decompose into two mercury atoms and one oxygen molecule, consisting of two atoms

GLOSSARY

Chemical equation: A representation of a chemical reaction in which the correct chemical formulas of the substances are written

Chemical reaction: A process which results in the formation of substances that were not present before the process occurred

Combination: A type of chemical change in which substances combine to form a different substance

Combustion: The reaction of a substance with oxygen in the atmosphere

Decomposition: A type of chemical change in which a compound is decomposed into its constituents

Displacement: A type of chemical change in which one element displaces another from its compound

Electrolysis: The decomposition of a compound by means of passing an electric current through the compound

Exchange: A type of chemical change in which the elements of one compound are exchanged with the elements of another compound

Products: The substances formed as a result of a chemical change (the chemical formulas of products appear on the right-hand side of the arrow in a chemical equation)

Reactants: The substances present before a chemical change occurs (the chemical formulas of reactants appear on the left-hand side of the arrow in a chemical equation)

SELF TEST

1. The chemist's shorthand describing a chemical reaction is called a(n) _____.

2. Which of the following elements exist as diatomic molecules when not combined with another element?
   (a) Aluminum
   (b) Nitrogen
   (c) Oxygen
   (d) Argon
   (e) Chlorine

3. Balance the following chemical equations, and classify each as either combination, decomposition, displacement, or exchange:
   (a) $PbO \xrightarrow{\Delta} Pb + O_2$
   (b) $Zn + O_2 \rightarrow ZnO$
   (c) $Mg + H(C_2H_3O_2) \rightarrow H_2 + Mg(C_2H_3O_2)_2$
   (d) $Cu + Ag(NO_3) \rightarrow Cu(NO_3)_2 + Ag$
   (e) $Cd(NO_3)_2 + (NH_4)_2S \rightarrow CdS + NH_4(NO_3)$

4. Write the correct formula for the missing product or reactant in each of the following equations:
   (a) $Zn + CuSO_4 \rightarrow Cu + \underline{\qquad}$
   (b) $Mg + 2HCl \rightarrow MgCl_2 + \underline{\qquad}$
   (c) $Zn + H_2SO_4 \rightarrow H_2 + \underline{\qquad}$
   (d) $2NaCl + \underline{\qquad} \rightarrow PbCl_2 + 2NaNO_3$
   (e) $2Li + Cl_2 \rightarrow 2\underline{\qquad}$

5. Write correct balanced chemical equations for the chemical reactions occurring between the following pairs of substances:
   (a) Magnesium reacts with oxygen.
   (b) Sodium reacts with sulfur.
   (c) Oxygen reacts with acetylene (C_2H_2).
   (d) Lithium reacts with bromine.
   (e) Lead(II) nitrate reacts with sodium sulfide.

ANSWERS

1. (a) Chemical equation

2. (b) Nitrogen, (c) oxygen, and (e) chlorine

3. (a) $2PbO \xrightarrow{\Delta} 2Pb + O_2$ (decomposition)
   (b) $2Zn + O_2 \rightarrow 2ZnO$ (combination)
   (c) $Mg + 2H(C_2H_3O_2) \rightarrow H_2 + Mg(C_2H_3O_2)_2$ (displacement)
   (d) $Cu + 2Ag(NO_3) \rightarrow Cu(NO_3)_2 + 2Ag$ (displacement)
   (e) $Cd(NO_3)_2 + (NH_4)_2S \rightarrow CdS + 2(NH_4)(NO_3)$ (exchange)

4. (a) $Zn(SO_4)$
   (b) H_2
   (c) $Zn(SO_4)$
   (d) $Pb(NO_3)_2$
   (e) $LiCl$

5. (a) $2Mg + O_2 \rightarrow 2MgO$
   (b) $2Na + S \rightarrow Na_2S$
   (c) $2C_2H_2 + 5O_2 \rightarrow 4CO_2 + 2H_2O$
   (d) $2Li + Br_2 \rightarrow 2LiBr$
   (e) $Pb(NO_3)_2 + Na_2S \rightarrow PbS + 2Na(NO_3)$

EXERCISES

1. What is a chemical reaction?

2. Draw diagrams representing molecules involved in the combustion of methane. (See Figs. 7-2 and 7-3.)

3. Name five simple types of reactions.

4. Describe the process that occurs in each of the following chemical reactions:
   (a) Combination
   (b) Decomposition
   (c) Displacement
   (d) Exchange
   (e) Combustion

5. For each of the following examples, classify the type of chemical reaction that will occur.
   (a) Magnesium reacts with sulfur.
   (b) Calcium reacts with chlorine.
   (c) Mercury(II) oxide is heated.
   (d) Calcium carbonate is heated.
   (e) Tin reacts with hydrogen sulfate.
   (f) Magnesium reacts with copper(II) nitrate.
   (g) Silver nitrate reacts with sodium chloride.
   (h) Sodium sulfide reacts with cadmium nitrate.
   (i) Octane (C_8H_{18}) reacts with oxygen.
   (j) Ethyl alcohol (C_2H_6O) reacts with oxygen.

6 Fill in the correct chemical formula for the missing reactant or product:
   (a) $2Na + $ _____ $\rightarrow 2NaCl$
   (b) $4Al + 3$ _____ $\rightarrow 2Al_2O_3$
   (c) $Mg + 2$ _____ $\rightarrow H_2 + MgCl_2$
   (d) $H_2(SO_4) + Zn \rightarrow H_2 + $ _____
   (e) $2NaCl + Pb(NO_3)_2 \rightarrow 2$ _____ $+ PbCl_2$
   (f) $C_2H_6 + \frac{7}{2}$ _____ $\rightarrow 2CO_2 + 3H_2O$
   (g) $CaCl_2 + $ _____ $\rightarrow 2NaCl + Ca(CO_3)$
   (h) $Mg + $ _____ $\rightarrow Cu + Mg(NO_3)_2$
   (i) $Zn + Pb(NO_3)_2 \rightarrow Pb + $ _____
   (j) $Cu + 2Ag(NO_3) \rightarrow 2$ _____ $+ Cu(NO_3)_2$

7 Translate each of the following word equations into *unbalanced* chemical equations. (Be sure to write the *correct formulas*!)
   (a) Hydrogen chloride (hydrochloric acid) reacts with sodium hydroxide to yield water and sodium chloride.
   (b) Iron reacts with hydrogen sulfate (sulfuric acid) to yield hydrogen gas and iron(II) sulfate.
   (c) Sodium metal reacts with water to yield hydrogen gas and sodium hydroxide.
   (d) Sodium bromide reacts with lead(II) nitrate to yield lead(II) bromide and sodium nitrate.
   (e) Magnesium metal reacts with bromine gas to yield magnesium bromide.

(f) Ferric hydroxide reacts with hydrogen bromide to yield water and ferric bromide.
(g) Propane (C_3H_8) reacts with oxygen to yield carbon dioxide and water.
(h) Carbon reacts with oxygen to yield carbon monoxide.
(i) Hydrogen peroxide decomposes to yield water and oxygen.

8 Rewrite and balance each of the chemical equations you obtained in Exercise 7.

9 Write the correct chemical formulas of the products formed for each of the following chemical reactions:
(a) $Al + O_2 \rightarrow$ _____
(b) $K_2S + Pb(NO_3)_2 \rightarrow$ _____ + _____
(c) $C_2H_6 + O_2 \rightarrow$ _____ + _____
(d) $H_3(PO_4) + Ca(OH)_2 \rightarrow$ _____ + _____
(e) $Zn(CO_3) \xrightarrow{\Delta} ZnO +$ _____
(f) $H(NO_3) + Al(OH)_3 \rightarrow$ _____ + _____
(g) $Ca + O_2 \rightarrow$ _____
(h) $N_2 + H_2 \rightarrow$ _____
(i) $Zn + H(C_2H_3O_2) \rightarrow$ _____ + _____
(j) $Hg(NO_3)_2 + (NH_4)_2S \rightarrow$ _____ + _____

10 Balance each of the equations in Exercise 9.

11 Write correct, complete, and balanced equations for each of the following chemical reactions:
(a) Sodium reacts with chlorine.
(b) Magnesium hydroxide reacts with hydrogen sulfate.
(c) Calcium metal reacts with hydrogen chloride.
(d) Tin reacts with hydrogen acetate.
(e) Methane reacts with oxygen.
(f) Hydrogen chloride reacts with aluminum carbonate.
(g) Calcium chloride reacts with sodium sulfate.
(h) Copper(II) nitrate reacts with potassium hydroxide.
(i) Copper reacts with sulfur.
(j) Copper reacts with silver nitrate.

MATHEMATICS AND CHEMISTRY II

MATHEMATICS AND CHEMISTRY II

CHAPTER TOPICS

In this chapter you will use nearly all the chemical concepts you have learned thus far to carry out calculations concerning chemical equations. Notice, again, how building on previously learned concepts is so important in mastering the study of chemistry. The most important operations and concepts to be learned from this chapter are how to

Interpret a balanced chemical equation in terms of *molecules*

Interpret a balanced chemical equation in terms of *moles*

Calculate the number of moles of one substance (required or produced) in a balanced chemical equation when given the number of moles of another substance

Calculate the number of grams of one substance (required or produced) in a balanced chemical equation when given the number of grams of another substance

8-1 CALCULATIONS BASED ON CHEMICAL EQUATIONS

The chemical equations studied in Chap. 7 may be interpreted in several different ways. They may be looked upon as a shorthand for the word description of the chemical reaction, or they may be considered as *quantitative* descriptions. Again, as a quantitative description, several different interpretations may be applied. For example, let us consider the *balanced equation* for the complete combustion of methane:

$$CH_4 + 2O_2 \rightarrow CO_2 + 2H_2O$$

This equation may be interpreted as one molecule of methane reacts with two molecules of oxygen to yield one molecule of carbon dioxide plus two molecules of water, thus establishing the *molecular ratio* for this reaction as 1:2:1:2. In other words, for every one molecule of methane that reacts, two molecules of oxygen are required, one molecule of carbon dioxide is formed, and two molecules of water are formed. If two molecules of methane react, four molecules of oxygen are required, and so on.

EXAMPLE

Calculate the number of molecules of oxygen needed to react with 1,000 molecules of methane:

$$(1 \times 10^3 \text{ molecules of CH}_4) \left(\frac{2 \text{ molecules of O}_2}{1 \text{ molecule of CH}_4} \right) = 2 \times 10^3 \text{ molecules of O}_2$$

Note that the conversion factor, two molecules O_2/1 molecule CH_4, is obtained from the *coefficients* of the equation. This emphasizes the importance of balancing the equation correctly before attempting *any* quantitative calculations based on the equation. Note the use of *unit analysis*.

Since the mole represents *a fixed number of molecules,* the *molecular ratio* obtained from a balanced equation is also the ratio of *moles* involved in the process.

EXAMPLE How many moles of O_2 are required to completely burn 0.25 mole of methane?

$$(0.25 \; \cancel{\text{mole of CH}_4}) \left(\frac{2 \text{ moles of } O_2}{1 \; \cancel{\text{mole of CH}_4}} \right) = 0.50 \text{ mole of } O_2$$

Again, note that the conversion factor

$$\frac{2 \text{ moles of } O_2}{1 \text{ mole of } CH_4}$$

comes from the balanced equation

$$CH_4 + 2O_2 \rightarrow CO_2 + 2H_2O$$
$$\uparrow \qquad\qquad\qquad \uparrow$$
$$(1 \text{ mole}) \qquad\quad (2 \text{ moles})$$

Note the use of unit analysis.

In Chap. 6 we learned that 1 mole of a substance weighs the number of grams represented by the molecular weight; therefore we can convert moles into grams. This permits us to establish the weights of the substances represented by an equation.

How many grams of oxygen are needed to completely burn 0.75 mole of methane? **EXAMPLE**

The balanced chemical equation is

$$CH_4 + 2O_2 \rightarrow CO_2 + 2H_2O$$

$$(0.75 \text{ mole of CH}_4) \left(\frac{2 \text{ moles of } O_2}{1 \text{ mole of CH}_4} \right) = 1.5 \text{ moles of } O_2$$

Then

$$(1.5 \text{ moles of } O_2) \left(\frac{32.0 \text{ g of } O_2}{1 \text{ mole of } O_2} \right) = 48.0 \text{ g of } O_2$$

or, in a single equation,

$$(0.75 \text{ mole CH}_4) \left(\frac{2 \text{ moles of } O_2}{1 \text{ mole of CH}_4} \right) \left(\frac{32.0 \text{ g of } O_2}{1 \text{ mole of } O_2} \right) = 48.0 \text{ g of } O_2$$

Again, note that the conversion factor

$$\frac{2 \text{ moles of } O_2}{1 \text{ mole of } CH_4}$$

is obtained from the balanced chemical equation for the reaction.

The type of problem that is most frequently encountered in practice is the calculation of the weights of different substances involved in chemical changes.

EXAMPLE Calculate the weight of oxygen needed to completely burn 500 g of methane.

The balanced equation is

$$CH_4 + 2O_2 \rightarrow CO_2 + 2H_2O$$

$$(500 \cancel{\text{ g of } CH_4}) \left(\frac{1 \cancel{\text{ mole of } CH_4}}{16.0 \cancel{\text{ g of } CH_4}}\right) \left(\frac{2 \cancel{\text{ moles of } O_2}}{1 \cancel{\text{ mole of } CH_4}}\right) \left(\frac{32.0 \text{ g of } O_2}{1 \cancel{\text{ mole of } O_2}}\right) = 2{,}000 \text{ g of } O_2$$

↑
(From balanced equation)

As in most chemical calculations there is a certain pattern to this type of calculation. Here the pattern is:

1   Write a complete and balanced chemical equation.

2   Convert grams of a given substance to moles of that substance.

3   Convert moles of a given substance to moles of the desired substance using the coefficients in the balanced equation and unit analysis.

4   Convert moles of the desired substance to grams of the desired substance.

Let us consider an example to illustrate the significance of this type of problem.

EXAMPLE

Suppose that a mountain cabin having dimensions of 10 ft by 12 ft by 8 ft was completely buried by snow so that there was no air circulation. Also suppose that a natural-gas heater in the cabin is burning methane at the rate of 300 g/hr. Assuming complete combustion, how long would it take for the oxygen in the cabin to be completely consumed?

Step 1   Calculate the volume of oxygen gas available:

$$10 \text{ ft} \times 12 \text{ ft} \times 8 \text{ ft} = 960 \text{ ft}^3 \text{ of air}$$

This is about 20% oxygen, and so we have 960 ft^3 × 0.20 = 192 ft^3 of oxygen in the cabin.

Step 2 Convert cubic feet to liters:

$$192 \text{ ft}^3 \times \left(\frac{28.3 \text{ liters}}{1 \text{ ft}^3}\right) = 5.4 \times 10^3 \text{ liters of } O_2$$

Step 3 Calculate the moles of oxygen available:

There are 22.4 liters/mole of any gas at 0°C and 760 mm pressure (this will be discussed more fully later). Therefore,

$$(5.4 \times 10^3 \text{ liters}) \left(\frac{1 \text{ mole}}{22.4 \text{ liters}}\right) = 2.4 \times 10^2 \text{ moles of } O_2 \text{ available}$$

Step 4 Find the time to consume all of the O_2:

$$\left(\frac{16.0 \text{ g of } CH_4}{1 \text{ mole of } CH_4}\right)\left(\frac{1 \text{ mole of } CH_4}{2 \text{ moles of } O_2}\right)\left(\frac{2.4 \times 10^2 \text{ moles of } O}{1}\right)\left(\frac{1 \text{ hr}}{3.0 \times 10^2 \text{ g of } CH_4}\right)$$

$$= 6.4 \text{ hr}$$

This example is a bit more involved than the problems in the exercises. More experience at problem solving will facilitate the solution of this type of problem (concerning the calculation of the weights of substances involved in chemical changes).

EXAMPLE Calculate the number of molecules of oxygen needed to react with 4 g of methane. (This type of example contributes to an understanding of the process but is not usually a practical problem.)

The balanced equation is

$$CH_4 + 2O_2 \rightarrow CO_2 + 2H_2O$$

$$(4 \text{ g of CH}_4) \left(\frac{1 \text{ mole of CH}_4}{16.0 \text{ g of CH}_4} \right) \left(\frac{2 \text{ moles of O}_2}{1 \text{ mole of CH}_4} \right) \left(\frac{6.02 \times 10^{23} \text{ molecules}}{1 \text{ mole}} \right)$$

(From the molecular weight)   (From the equation)   (From Avogadro's number)

$$= 3.01 \times 10^{23} \text{ molecules}$$

EXAMPLE Ammonia (NH_3), which is used in cleansers and in fertilizers, can be made from nitrogen gas and hydrogen gas. The balanced equation for the chemical reaction is

$$N_2 + 3H_2 \rightarrow 2NH_3$$

Part a   How many moles of NH_3 can be made by the complete reaction of 3.5 moles of hydrogen gas? Since the balanced equation shows that the ratio of molecules of NH_3 formed to H_2 used is 2:3, the desired mole ratio is also 2:3:

$$\text{Moles of } NH_3 = \left(\frac{2 \text{ moles of } NH_3}{3 \text{ moles of } H_2}\right)(3.5 \text{ moles of } H_2) = 2.3 \text{ moles}$$

Part b How many grams of NH_3 can be generated from 0.300 mole of nitrogen gas?

$$\text{Moles of } NH_3 = (0.300 \text{ mole of } N_2)\left(\frac{2 \text{ moles of } NH_3}{1 \text{ mole of } N_2}\right) = 0.600 \text{ mole}$$

Now, 0.600 mole of NH_3 can be converted to grams of NH_3 since we can calculate that 1.00 mole of $NH_3 = 17.0$ g (14.0 awu + 3.03 awu = 17.0 awu):

$$\text{Grams of } NH_3 = \left(\frac{17.0 \text{ g}}{1 \text{ mole}}\right)(0.600 \text{ mole}) = 10.2 \text{ g}$$

Alternatively, the problem could be set up in one step:

$$(0.300 \text{ mole of } N_2)\left(\frac{2 \text{ moles of } NH_3}{1 \text{ mole of } N_2}\right)\left(\frac{17.0 \text{ g}}{1 \text{ mole of } NH_3}\right) = 10.2 \text{ g}$$

Part c How many grams of nitrogen would be required to produce 35.0 g of NH_3?

$$\text{Moles of } NH_3 = (35.0 \text{ g}) \left(\frac{1 \text{ mole}}{17.0 \text{ g}}\right) = 2.06 \text{ moles}$$

$$\text{Moles of } N_2 = (2.06 \text{ moles of } NH_3) \left(\frac{1 \text{ mole of } N_2}{2 \text{ moles of } NH_3}\right) = 1.03 \text{ moles}$$

$$\text{Grams of } N_2 = (1.03 \text{ moles of } N_2) \left(\frac{28.0 \text{ g}}{1 \text{ mole of } N_2}\right) = 28.9 \text{ g}$$

or, in one step,

$$\text{Grams of } N_2 =$$
$$(35.0 \text{ g of } NH_3) \left(\frac{1 \text{ mole of } NH_3}{17.0 \text{ g of } NH_3}\right) \left(\frac{1 \text{ mole of } N_2}{2 \text{ moles of } NH_3}\right) \left(\frac{28.0 \text{ g of } N_2}{1.00 \text{ mole of } N_2}\right)$$
$$= 28.9 \text{ g}$$

Another way of solving the previous example (as well as earlier examples in the chapter) involves the so-called "ratio-proportion" method. Rewriting the equation for the reaction, $N_2 + 3H_2 \rightarrow 2NH_3$.

The weight of 1 mole of N_2 is 28.0 g, and the weight of 2 moles of NH_3 is 2.0×17.0 g, or 34.0 g; therefore, the ratio of the weight of N_2 to the weight of NH_3 is 28.0:34.0. A proportion may be set up as

$$\frac{X \text{ g of } N_2}{28.0 \text{ g of } N_2} = \frac{35.0 \text{ g of } NH_3}{2 \times 17.0 \text{ g of } NH_3}$$

Solving the above for X gives

$$X = \left(\frac{35.0 \text{ g of } NH_3}{2 \times 17.0 \text{ g of } NH_3}\right)(28.0 \text{ g of } N_2) = 28.9 \text{ g } N_2$$

Or it could be set up as

$$\begin{array}{cc} X & 35.0 \text{ g} \\ N_2 + 3H_2 \rightarrow & 2NH_3 \\ 28.0 \text{ awu} & 34.0 \text{ awu} \end{array}$$

or

$$\frac{X}{28.0 \text{ awu}} = \frac{35.0 \text{ g}}{34.0 \text{ awu}}$$

Then the problem may be solved for X.

You will observe that the *mole method* of calculating the number of grams of N_2 needed to form 35.0 g of NH_3 (see Part c in the previous example) and the *ratio-proportion method* of calculating the number of grams of N_2 needed to form 35.0 g NH_3 are really the *same calculation*. Thus from Part c (the mole method):

Grams of $N_2 =$

$$(35.0 \text{ g of NH}_3) \left(\frac{1 \text{ mole of NH}_3}{17.0 \text{ g of NH}_3}\right) \left(\frac{1 \text{ mole of N}_2}{2 \text{ moles of NH}_3}\right) \left(\frac{28.0 \text{ g of N}_2}{1.00 \text{ mole of N}_2}\right)$$

$$= 28.9 \text{ g of N}_2$$

and, from the ratio-proportion method,

$$X = \left(\frac{35.0 \text{ g}}{1}\right) \left(\frac{1}{17.0 \text{ g} \times 2}\right) \left(\frac{28.0 \text{ g}}{1}\right) = 28.9 \text{ g of N}_2$$

All the problems discussed in this chapter thus far have included an implied assumption. This assumption is that the necessary amount of material to react with the given quantity *was available*. In many practical situations this is not true. When two reactants (or *reagents*) come together in quantities *different* than the quantities represented by a balanced equation, the reaction will continue until one of the two reagents is all consumed; then the reaction will stop. The substance that is not all consumed remains unreacted and is said to be present in *excess*. The substance that is all consumed is said to be the *limiting reagent* because it *limits* the amount of product that can be formed.

EXAMPLE If two tablets of milk of magnesia containing 10.0 g of magnesium hydroxide are swallowed and contact a human stomach containing 3.65 g of hydrochloric acid, how many grams of magnesium chloride will be formed?

Step 1 The balanced equation for the reaction will be

$$Mg(OH)_2 + 2HCl \rightarrow MgCl_2 + 2H(OH)$$

Step 2 The molecular weight of $Mg(OH)_2$ is

$$\begin{array}{r} 24.3 \\ 32.0 \\ 2.0 \\ \hline 58.3 \text{ g/mole} \end{array}$$

Therefore we have available

$$\frac{10.0 \text{ g}}{58.3 \text{ g/mole}} = 0.171 \text{ mole of } Mg(OH)_2$$

The molecular weight of HCl is

$$\begin{array}{r} 1.0 \\ 35.5 \\ \hline 36.5 \text{ g/mole} \end{array}$$

Therefore we have available

$$\frac{3.65 \text{ g}}{36.5 \text{ g/mole}} = 0.100 \text{ mole of HCl}$$

Since the ratio of molecules in the equation is 1:2, this reaction will require one-half as many moles of $Mg(OH)_2$ as moles of HCl.

Since one-half of 0.100 mole is 0.050 mole, only 0.050 mole of the $Mg(OH)_2$ can react with HCl; the remainder of the $Mg(OH)_2$ (or $0.171 - 0.050 = 0.121$ mole) will be in *excess*, i.e., unreacted. Thus HCl is the *limiting reagent* in this process.

Step 3 The number of moles of $MgCl_2$ formed will be

$$\text{Moles of } MgCl_2 = \left(\frac{0.100 \text{ mole of HCl}}{1}\right) \left(\frac{1 \text{ mole of } MgCl_2}{2 \text{ moles of HCl}}\right)$$

$$= 0.050 \text{ mole} \qquad \text{(From the equation)}$$

Step 4 Since the molecular weight of $MgCl_2$ is 95.3, the number of grams of $MgCl_2$ formed will be

$$\text{Grams of } MgCl_2 = \left(\frac{0.050 \text{ mole}}{1}\right) \left(\frac{95.3 \text{ g}}{\text{mole}}\right) = 4.76 \text{ g}$$

TABLE 8-1 Interpretation of a chemical equation

$2H_2$	+	O_2	→	$2H_2O$
Two molecules of hydrogen	react with	one molecule of oxygen	to yield	two molecules of water
Two moles of hydrogen	react with	1 mole of oxygen	to yield	2 moles of water
4.04 g of hydrogen	react with	32.0 g of oxygen	to yield	36.0 g of water
4.04 tons of hydrogen	react with	32.0 tons of oxygen	to yield	36.0 tons of water

In summary, consider Table 8-1, which shows the different ways a chemical equation may be interpreted.

GLOSSARY[1]

Balancing coefficients: The number appearing *in front of* the chemical formula in a balanced chemical equation

Excess reagent: The substance that is "left over" partially unreacted at the end of a reaction

Limiting reagent: The substance that limits the amount of product formed

---

[1]The glossary for this chapter is quite short because most of the terms have been introduced in previous chapters.

Mole ratio: A ratio that is numerically equal to the molecular ratio

Molecular ratio: The ratio in which molecules react in a chemical change (the same as the ratio of balancing coefficients in a balanced chemical equation)

1   Copper metal can be dissolved with nitric acid. The chemical equation for the reaction which occurs is

SELF TEST

$$Cu + 4HNO_3 \rightarrow Cu(NO_3)_2 + 2NO_2 + 2H_2O$$

Given the above equation, fill in the following blanks:

One atom of copper will react with (a) _____ molecule(s) of hydrogen nitrate to form (b) _____ molecule(s) of copper(II) nitrate, (c) _____ molecule(s) of nitrogen dioxide, and (d) _____ molecule(s) of water. For complete reaction, 0.50 mole of hydrogen nitrate reacts with (e) _____ mole(s) of copper metal to form (f) _____ mole(s) of copper(II) nitrate, (g) _____ mole(s) of nitrogen dioxide, and (h) _____ mole(s) of water. In order to produce 3.0 moles of NO_2, (i) _____ mole(s) of Cu must react with (j) _____ mole(s) of HNO_3; (k) _____ mole(s) of $Cu(NO_3)_2$ and (l) _____ mole(s) of H_2O will also be generated.

2 Fluorine gas, a very corrosive and toxic substance, can be generated by decomposing hot, liquid potassium fluoride according to the chemical equation

$$2KF \rightarrow F_2 + 2K$$

(a) How many grams of F_2 could be generated from 10.0 g of KF?
(b) How many grams of K could be formed from 10.0 g of KF?

ANSWERS

1 (a) 4 molecules of HNO_3 (g) 0.250 mole of NO_2
   (b) 1 molecule of $Cu(NO_3)_2$ (h) 0.250 mole of H_2O
   (c) 2 molecules of NO_2 (i) 1.5 moles of Cu
   (d) 2 molecules of H_2O (j) 6.0 moles of HNO_3
   (e) 0.125 mole of Cu (k) 1.5 moles of $Cu(NO_3)_2$
   (f) 0.125 mole of $Cu(NO_3)_2$ (l) 3.0 moles of H_2O

2 (a) 3.28 g of F_2
   (b) 6.75 g of K

EXERCISES

1 (a) Write a balanced equation for the complete reaction between hydrogen sulfate and potassium hydroxide to form potassium sulfate plus water.

(b) How many molecules of each substance are represented in the balanced chemical equation?
(c) How many moles of potassium hydroxide are required to completely react with 1 mole of hydrogen sulfate?
(d) Calculate the quantities needed to complete the following sentence:
_____ mole(s) of potassium hydroxide are required to completely react with 0.25 mole of hydrogen sulfate to yield _____ mole(s) of potassium sulfate and _____ mole(s) of water.

2. (a) Calculate the number of moles of carbon dioxide, and the number of moles of water that can be formed from the complete combustion of 0.25 mole of methane.
   (b) Calculate the number of grams of carbon dioxide, and the number of grams of water that will be produced by the complete combustion of 0.25 mole of methane.

3. Calculate the number of molecules of carbon dioxide and the number of molecules of water that will be formed in the complete combustion of 1,000 molecules of methane.

4. Calculate the number of grams of water and the number of grams of carbon dioxide that will be formed in the complete combustion of 500 g of methane.

5. Given the balanced equation $H_2 + Cl_2 \rightarrow 2HCl$:

(a) How many molecules of Cl_2 would be needed to form 50 molecules of HCl?

(b) What would be the weight (in atomic weight units) of the number of Cl_2 molecules calculated in part (a)?

(c) How many molecules of H_2 would be needed to form 50 molecules of HCl?

(d) What is the weight ratio of 50 molecules of Cl_2 to 50 molecules of H_2?

(e) What is the ratio of the molecular weight of Cl_2 to the molecular weight of H_2?

6. The complete combustion of butane gas (C_4H_{10}), which is used in butane cigarette lighters, is

$$2C_4H_{10} + 13O_2 \rightarrow 8CO_2 + 10H_2O$$

Given this equation, fill in the following blanks.

Three moles of C_4H_{10} requires _____ mole(s) of O_2 for complete reaction to form _____ mole(s) of CO_2 and _____ mole(s) of H_2O. One-half mole of O_2 will completely react with _____ mole(s) of C_4H_{10} to form _____ mole(s) of CO_2 and _____ mole(s) of H_2O. _____ g of O_2 would be required to react with 10.0 g of C_4H_{10} to form _____ mole(s) of CO_2 and _____ mole(s) of H_2O.

7   Mercury metal can be produced by heating the ore called *cinnabar*. The chemical equation describing this reaction is

   $2HgO \rightarrow 2Hg + O_2$

   Calculate the number of grams of mercury metal which could be produced from the complete reaction of 50.0 g of mercuric oxide.

8   Write the equation for the reaction of aluminum with hydrochloric acid. Calculate, on the basis of this equation, the
   (a) Number of moles of HCl needed to react with 0.45 mole of aluminum
   (b) Weight of HCl needed to react with 0.45 mole of aluminum
   (c) Number of moles of aluminum chloride that can be formed from 0.45 mole of aluminum
   (d) Weight of aluminum chloride that can be formed from 0.45 mole of aluminum

9   Write the equation for the reaction of calcium with chlorine. Calculate, on the basis of this equation, the
   (a) Relative weights of calcium, chlorine, and calcium chloride involved in this process
   (b) Weight of chlorine needed to combine with 10 g of calcium
   (c) Weight of calcium chloride that can be formed from 10 g of chlorine
   (d) Weight of calcium chloride that can be formed if 10 g of calcium and 10 g of chlorine are available (the answer is not "20 g")

10. Write the equation for the reaction of potassium with water. Calculate, on the basis of this equation, the
    (a) Number of moles of hydrogen that can be formed, starting with 0.30 mole of potassium
    (b) Number of grams of hydrogen that can be formed, starting with 0.30 mole of potassium
    (c) Number of moles of water that will react with 0.30 mole of potassium
    (d) Number of grams of water that will react with 0.30 mole of potassium

11. Quicklime, CaO, will react with water to form slaked lime, $Ca(OH)_2$.
    (a) Write a balanced equation for this process.
    (b) Calculate, on the basis of this equation, the weight of slaked lime that can be formed from 1 ton of quicklime.

12. Given the equation $MnO_2 + 4HCl \rightarrow Cl_2 + MnCl_2 + 2H_2O$, calculate, on the basis of this equation, the
    (a) Number of moles of free chlorine that can be formed from 0.40 mole of HCl
    (b) Number of grams of free chlorine that can be produced from 0.40 mole of HCl
    (c) Number of grams of free chlorine that can be produced from 14.6 g of HCl
    (d) Explain why the answer to part (b) is the same as the answer to part (c).

13 Consider the reaction $N_2 + 3H_2 \rightarrow 2NH_3$. Fill in the missing factors below if 20.0 g of NH_3 are formed:

(a) Moles NH_3 = (20.0 g of NH_3) (_____)

(b) Moles H_2 necessary =
(20.0 g of NH_3) (_____) (_____)

(c) Grams H_2 necessary =
(20.0 g of NH_3) $\left(\dfrac{1 \text{ mole of } NH_3}{17.0 \text{ g of } NH_3}\right)$ (_____) (_____)

(d) Moles N_2 necessary =
(20.0 g of NH_3) $\left(\dfrac{1 \text{ mole of } NH_3}{17.0 \text{ g of } NH_3}\right)$ (_____)

(e) Grams N_2 necessary =
(20.0 g of NH_3) $\left(\dfrac{1 \text{ mole of } NH_3}{17.0 \text{ g of } NH_3}\right)$ (_____) (_____)

14 The complete combustion of butane (C_4H_{10}) can be represented by the balanced chemical equation

$$2C_4H_{10} + 13O_2 \rightarrow 10H_2O + 8CO_2$$

Fill in the following blanks:

For complete combustion of 1 mole of butane, _____ mole(s) of O_2 will be needed, and _____ mole(s) of H_2O along with _____ mole(s) of CO_2 will be formed. One-third of a mole of O_2 will completely react with _____ mole(s) of butane to yield _____ mole(s) of H_2O and _____ mole(s) of CO_2. 22.0 g of CO_2 and _____ g of H_2O will be formed from the complete reaction of _____ g of O_2 and _____ g of butane.

15 One of the chemical reactions which occurs when copper ore is smelted can be represented by

$$2CuS + 3O_2 \rightarrow 2CuO + 2SO_2 \qquad (i)$$

(a) How many grams of sulfur dioxide gas would be generated by the complete reaction of 2,000 kg of CuS?

(b) The sulfur dioxide can be converted to sulfuric acid (H_2SO_4) by the following reactions:

$$2SO_2 + O_2 \rightarrow 2SO_3 \qquad (ii)$$

$$SO_3 + H_2O \rightarrow H_2SO_4 \qquad (iii)$$

How many grams of H_2SO_4 could be made from smelting 2,000 kg of

CuS? [You should use the sequence of equations (i) to (iii) to solve the problem.]

16 In Los Angeles County during 1969 approximately 8 million gal of gasoline were burned per day by motor vehicles (8 million gal of gasoline weighs about 22.0×10^6 kg).

(a) If we assume that the molecular formula of gasoline is C_8H_{18}* and that 80 percent of the gasoline is burned according to the equation

$$2C_8H_{18} + 25O_2 \rightarrow 16CO_2 + 18H_2O$$

calculate the number of kilograms of CO_2 generated per day by the gasoline-powered vehicles.

(b) If we assume that 20 percent of the 22.0×10^6 kg of gasoline burns incompletely according to the equation

$$2C_8H_{18} + 17O_2 \rightarrow 16CO + 18H_2O$$

calculate the weight of carbon monoxide formed each day by gasoline-powered vehicles.

17 Carbon monoxide gas is toxic. When the concentration is 1,000 ppm† in the air, it will produce unconsciousness in 1 hr and death in 4 hr. If a

*Gasoline is actually a complex mixture of compounds, which is varied seasonally.
†ppm means *parts per million*.

closed one-room apartment has dimensions 3 m × 3 m × 3 m, its volume is 27m^3. The weight of air contained in the room is

$$(2.7 \times 10^1 \text{ m}^3)(1.0 \times 10^6 \text{ cm}^3/\text{m}^3)(1 \text{ liter}/1 \times 10^3 \text{ cm}^3)(1.29 \text{ g/liter}) = 3.5 \times 10^4 \text{ g}$$

It would require $(3.5 \times 10^4 \text{ g})(1 \times 10^3/1 \times 10^6) = 3.5 \times 10^1$ g of CO to attain a concentration of 1,000 ppm of CO. Calculate the number of grams of carbon that will need to burn to produce a lethal concentration of carbon monoxide in the apartment, assuming that when charcoal burns, 10 percent of it reacts according to the equation

$$2C + O_2 \rightarrow 2CO$$

18 Sugar (glucose) will ferment according to the reaction

$$C_6H_{12}O_6 \rightarrow 2CO_2 + 2C_2H_6O$$

Calculate the number of grams of alcohol that can be formed from the fermentation of 30.0 g of glucose.

19 At a sufficiently high temperature such as occurs in the cylinders of an automobile engine during combustion, nitrogen reacts with oxygen to

form NO, contributing to air pollution. The equation may be written as

$$N_2 + O_2 \rightarrow 2NO$$

Calculate the
(a) Number of grams of NO that could be formed from 4.0 g of nitrogen
(b) Number of moles of NO that can be formed from 4.0 g of N_2

20   The balanced equation for the combustion of methane is

$$CH_4 + 2O_2 \rightarrow CO_2 + 2H_2O$$

(a) If 50.0 g of O_2 is mixed with 10.0 g of CH_4 and the mixture burns completely, which reactant is the limiting reagent?
(b) Which reactant is the excess reagent?
(c) How many moles of CO_2 are formed?
(d) How many moles of H_2O are formed?
(e) How many grams of CO_2 are formed?
(f) How many grams of H_2O are formed?
(g) How many grams of the limiting reagent remain unreacted?

21   If baking soda, $Na(HCO_3)$, is added to vinegar, a chemical reaction

occurs which forms CO_2. The balanced chemical equation is

$$Na(HCO_3) + H(C_2H_3O_2) \rightarrow CO_2 + H_2O + Na(C_2H_3O_2)$$

(a) If 21.0 g of sodium bicarbonate is placed in vinegar containing 30.0 g of $H(C_2H_3O_2)$, which reactant is the limiting reagent?
(b) Which reactant is the excess reagent?
(c) How many moles of CO_2 are formed?
(d) How many grams of CO_2 are formed?
(e) How many moles of H_2O are formed?
(f) How many grams of H_2O are formed?
(g) How many grams of the excess reagent remain unreacted?

SOLIDS, LIQUIDS, AND GASES — THE KINETIC MOLECULAR THEORY

SOLIDS, LIQUIDS, AND GASES — THE KINETIC-MOLECULAR THEORY

CHAPTER TOPICS

Matter may exist in either the solid, liquid, or gaseous state. In studying this chapter you should learn

The nature of substances that are *solids*

The nature of substances that are *liquids*

The nature of substances that are *gases*

The similarities and differences of these three states of matter

The fundamental concepts of the *kinetic-molecular theory*

How the behavior of a gas can be explained using the kinetic-molecular theory

How the behavior of solids and liquids may be explained in terms of the kinetic-molecular theory

How changes of state may be described by the kinetic-molecular theory

9-1 SOLIDS

Water is the most abundant compound on the surface of the earth. Approximately 71 percent of the earth's surface is covered with water. The total volume of water on the earth is about 32 million miles^3, and the total weight of water is about 1.4×10^{18} tons. Water can exist in three states: the *solid* state, or ice; the *liquid* state, or water; and the *gaseous* state, or steam.

Let us consider the familiar solid state of water—ice. A piece of ice has a definite shape and resists attempts to change that shape. This resistance to change of shape is characteristic of all solids. In ice the molecules of water are arranged in definite positions with respect to each other. They are held in place by the forces of attraction between the molecules. Figure 9-1 illustrates the regular arrangement of water molecules in ice.

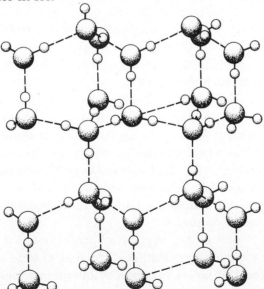

FIGURE 9-1 The structure of ice (*Adapted from Linus Pauling, "General Chemistry," 3d ed., W. H. Freeman and Company, 1970*)

FIGURE 9-2   The structure of sodium chloride

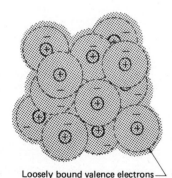

Loosely bound valence electrons

FIGURE 9-3   The structure of a metal

Many other solids consist of particles arranged in these definite patterns called *crystal lattices*. The structure of ice shown in Fig. 9-1 is an example of a crystal lattice in which the building units of the lattice are *molecules*. Sodium chloride is a solid composed of particles in which the building units are *ions*. The lattice structure of sodium chloride is illustrated in Fig. 9-2. Note that an individual sodium ion is *not* permanently attached to any specific chloride ion. Each sodium ion is surrounded by and electrostatically attracted to six nearest neighbor chloride ions. The structure of ice is an example of a crystal lattice composed of molecules. Here the attractive forces between the particles are much weaker than the attractive forces which exist between the charged ions in a solid like sodium chloride. The temperature at which a *molecular solid* (e.g., ice) melts is generally considerably lower than the temperature at which an *ionic solid* (e.g., sodium chloride) melts.

Another type of solid is the familiar *metallic solid* like iron or magnesium. The structure of a metallic solid is thought to be a regular arrangement of atoms which constitutes the crystal lattice, as shown in Fig. 9-3. This arrangement of atoms is held together by the valence electrons which are only loosely held by any particular atom and tend to be distributed over the entire piece of the metal. These valence electrons, which are so loosely held, can migrate within the piece of metal, and the migration of these electrons permits the easy conduction of an electric current.

Still another type of solid does not have its particles arranged in a regular pattern or crystal lattice. This type of solid is called an *amorphous solid*. Charcoal and tar are examples of amorphous solids, where the particles are distributed in a helter-skelter manner.

9-2 LIQUIDS

In a *liquid* the molecules are not held rigidly in a fixed position. The structure of a liquid is somewhat difficult to visualize. The molecules are rather free to move about and can roll about each other, but attractive forces between molecules still effectively hold the molecules near each other. A liquid is not a rigid material, as we know, but tends to assume the shape of any container in which it is placed. This is a consequence of the fact that the molecules are able to move about one another easily. However, the fact that the molecules are still quite close together is shown by the very small compressibility of a liquid. For example, the hydraulic braking system of an automobile makes use of the incompressibility of a liquid to transmit pressure equally to each of the brake shoes of the automobile.

The molecules of a liquid are attracted to each other, which causes a liquid to exhibit *surface tension*. Surface tension is what makes water droplets form, as seen when water sprayed on a waxed automobile "beads up." Small insects, like water skates, are supported by surface tension; and a small needle carefully placed on water can be supported by surface tension.

9-3 GASES

Most people are more familiar with the properties of liquids and solids than with those of gases even though we live in an "ocean" of gases called *air*. A *gas* is a substance which occupies the complete volume of a container. It is easy to show that a gas is a real substance which *occupies space*, however, by inverting a glass and pressing it into a large jar of water. If you do this, you will observe that the water is prevented from entirely filling the glass owing to the amount of air the glass contains.

We know that a gas *has weight,* too, since it is able to exert a force, as shown by the pressure of the wind on the sail of a boat. Also, remember that the atmosphere, which is composed of gases, exerts considerable *pressure* (or force per unit area) on the earth's surface. One significant difference between the properties of a gas and those of a liquid or solid is the great *compressibility* of a gas. For example, 72 ft^3 of air measured under normal atmospheric temperature and pressure may be compressed into a standard scuba diving tank, which has a volume of about 0.5 ft^3. This quality of great compressibility suggests that there is a great deal of empty space between the molecules which comprise a gas.

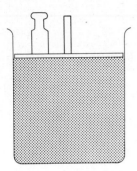

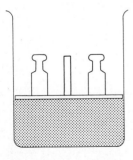

FIGURE 9-4 Behavior of a gas with a change in pressure

9-4 THE KINETIC-MOLECULAR THEORY

Let us describe the behavior of gases in more detail, and then we shall see how the kinetic-molecular theory can account for the observed behavior. A gas contained in a cylinder at constant temperature with a movable piston, as shown in Fig. 9-4, can be compressed by moving the piston down. Increasing the force on the piston causes the piston to move downward and decrease the volume of the gas. This behavior of a gas is summarized by *Boyle's law,* which states that at constant temperature the volume of a gas *decreases* with an *increase* in pressure. On the other hand, when the *pressure* exerted on a gas is kept constant, a different behavior is observed. This is summarized by *Charles' law,* which states that at constant pressure the volume of a gas will *increase* with an *increase* in the temperature.

From a careful study of the properties of the three states of matter, an imaginative model has been developed to help us visualize the particulate nature of a solid, liquid, or gas. This imaginative model with its related concepts is called the *kinetic-molecular theory*. The kinetic-molecular theory pictures a gas in the following way:

1. It is composed of tiny separate particles which we call *molecules*.
2. Each gas molecule travels in a straight line until it collides with something. When collisions occur, the molecules rebound with perfect elasticity; that is, there is no net loss of the energy of the molecules as a result of their collision.
3. The average speed at which a particular type of molecule will travel depends upon the temperature. The greater the temperature, the greater the speed of the molecules. The lower the temperature, the lower the speed. In fact, we might say that the temperature of a gas is caused by the average speed at which the molecules of that gas are traveling. The greater the average speed of the molecules, the higher the temperature of the substance.

Let us consider the way in which the relationship between pressure and volume of a gas may be explained in terms of the kinetic-molecular theory. (Refer to Fig. 9-5 in order to better understand the following discussion.) When a molecule of a gas collides with the wall of a container and rebounds, a force is exerted on the wall of that container. The forces resulting from the repeated

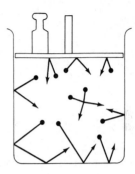

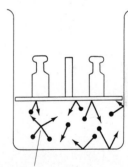

Two particles bouncing off each other

FIGURE 9-5 Behavior of a gas with a change in pressure showing possible molecular pathways.

collisions of gas molecules upon the walls of the container result in a steady pressure being exerted upon all the walls. Since the average speed at which any one type of gas molecule moves depends only upon the temperature, we can see that at a fixed temperature, the force a gas exerts upon the walls of a container will depend upon the frequency with which the molecules collide with the walls. For a given quantity of gas, this frequency will depend upon the volume which the gas occupies. If the volume is *decreased*, the same number of gas molecules moving at the same speed will be confined in a smaller volume. These molecules will collide with the walls of the container more frequently, resulting in an *increase* in pressure (as stated by Boyle's law). Similarly, if the volume of the container is increased, the gas molecules will collide with the walls of the container less frequently resulting in a *decrease* in pressure.

To help us visualize the application of the kinetic-molecular theory to the volume-temperature relationship of a gas, refer to the movable piston shown in Fig. 9-5. Let us assume the movable piston is capable of maintaining a constant pressure upon the gas that is enclosed. If the temperature of the gas is increased, the average speed of the molecules of the gas will be increased. This results in more frequent collisions with the walls of the container. Each collision of a molecule will exert a greater force upon the walls because the average speed of the molecule is greater. This means that the movable piston will be moved back, increasing the volume occupied by the gas (and therefore reducing the frequency of molecular collisions with the wall) to maintain the pressure at a constant value. We see that the volume will increase as the temperature increases at constant pressure (as stated by Charles' law).

Next let us consider the pressure-temperature relationship of a gas in terms of the kinetic-molecular theory. Visualize a situation where a given quantity of gas is contained within a fixed volume, such as a steel cylinder. As the temperature of the gas increases, the average speed of the molecules will increase. This will result in each collision of a molecule with a wall of the container exerting a greater force and also in more frequent collisions. We can easily see that the pressure exerted by the gas molecules will *increase* as the temperature is *increased*.

The properties of a *liquid* may also be interpreted in terms of the kinetic-molecular theory. At any temperature above $-273°C$ the molecules of a liquid are in motion. If the liquid is heated, the molecules move about each other more vigorously. At any given temperature a few molecules of the liquid are moving fast enough to overcome the forces of attraction between molecules and escape into the gaseous state; thus *evaporation* occurs. As the temperature increases, a higher percentage of the molecules develop this necessary amount of energy. In fact, *both* solids and liquids evaporate. The evaporation of a solid is called *sublimation*. Thus molecules are continuously escaping from either the solid or the liquid state. The escape of molecules of water from solid ice can be illustrated by the slow drying of frozen, freshly washed clothes on a clothesline. We are also familiar with the fact that the molecules of liquid water evaporate at room temperature. Since more molecules will be moving fast enough to break away from the solid or liquid at higher temperatures, the speed of evaporation *increases* with an *increase* in temperature.

When a liquid is heated to a certain temperature at atmospheric pressure, it is observed that the temperature of the liquid cannot be increased as long as the liquid is agitated. As heat is added, it is completely used up in vaporizing the liquid. This temperature is called the *boiling point*. The classical definition of the boiling point of a liquid is *the temperature at which the vapor pressure of the liquid is equal to the atmospheric pressure*. (This is another characteristic by which a substance can be recognized. The boiling point of water at sea level, for example, is 100°C.)

9-5 CHANGES OF STATE

Nearly all substances can exist in either the solid, liquid, or gaseous state under suitable conditions of temperature and pressure. We can illustrate the *changes of state* that can occur by using water as an example. Let us recall the structure of ice as shown in Fig. 9-1. Here the molecules are arranged in an orderly pattern, or lattice. As the temperature of the ice (or any *solid*) is increased, the molecules vibrate more vigorously about their fixed positions. When the amplitude of this vibration is large enough, the lattice is destroyed and melting occurs. This forms a *liquid* in which the molecules are more free to move. The temperature at which both the solid and liquid can coexist in contact with each other is called the *melting point*. When heat is added to a solid at its melting point, it is observed that no temperature change occurs during melting. Thus melting is an example of a change of state because a solid is changed into a liquid.

As the temperature of the liquid water is increased, the molecules move more rapidly and the rate of evaporation increases because a higher percentage of molecules are moving fast enough to escape from the surface of the liquid. These molecules escaping from the surface of the liquid produce a pressure called the *vapor pressure* of the liquid. When the vapor pressure of the liquid is equal to the applied atmospheric pressure, the liquid boils. The temperature at which this occurs is the *boiling point*. As heat is added to a liquid at its boiling point, the temperature remains constant during boiling.

GLOSSARY

Amorphous solid: A rigid substance whose molecules are not in an orderly spatial arrangement

Boiling point: The temperature at which the vapor pressure of the liquid is equal to the atmospheric pressure

Boyle's law: At constant temperature the volume of a gas *decreases* with an *increase* in pressure

Charles' law: At constant pressure the volume of a gas *increases* with an *increase* in temperature

Crystal lattice: The definite spatial arrangement of the particles which make up a solid

Evaporation: The escape of molecules from a liquid into the gaseous state

Gas: A substance which occupies the complete volume of a container, is very compressible at ordinary temperatures and pressures, has particles in random straight-line motion, and has distances between the molecules that are large compared to the size of the molecules themselves

Ionic solid: A solid whose crystal lattice building units are ions

Kinetic-molecular theory: See Sec. 9-4

Liquid: A substance that assumes the shape of its container, is not rigid, and has particles that are not in an orderly spatial arrangement

Melting point: The temperature at which both a liquid and solid can coexist in contact with each other

Metallic solid: A solid whose crystal lattice building units are atoms which have loosely bound valence electrons

Molecular solid: A solid whose crystal lattice building units are molecules

Pressure: The force per unit area (for example, $lb/in.^2$)

Solid: A rigid substance whose particles are in an orderly spatial arrangement

Sublimation: The escape of molecules from a solid into the gaseous state

Surface tension: The force which causes a liquid to form droplets produced by the forces of attraction between the particles at the surface of a liquid

Vapor pressure: The pressure exerted by molecules of a substance which have escaped from the solid or liquid state

SELF TEST

1. The three states of matter are _____, _____, and _____.

2. The particles of a _____ are spatially arranged in an orderly fashion.

3. The distances between the molecules of a _____ are large compared with the size of the molecules.

4. An orderly array of the building units of a solid is called a _____.

5. Iron is an example of a(n) _____ solid.

6. The force which causes water to form in droplets is called _____.

7   At constant pressure the volume of a gas _____ with an increase in temperature.

8   At constant temperature the volume of a gas _____ with an increase in pressure.

9   The escape of molecules from a solid to the gaseous state is called _____.

10 The temperature at which the solid substance and liquid substance can coexist is the _____.

ANSWERS
1   Solid, liquid, gas
2   Solid
3   Gas
4   Crystal lattice
5   Metallic
6   Surface tension
7   Increases
8   Decreases
9   Sublimation
10 Melting point

EXERCISES

1. What are the three states in which water may exist?

2. How do the physical properties of a gas differ from the physical properties of a liquid?

3. In which state of matter are molecules vibrating about fixed positions with respect to one another?

4. In which state of matter are the molecules held fairly close together but are still free to move about one another?

5. Which state of matter is the most compressible? Why?

6. State Charles' law.

7. State Boyle's law.

8. What are the principal postulates of the kinetic-molecular theory?

9. Explain Boyle's law in terms of the kinetic-molecular theory.

10. Explain Charles' law in terms of the kinetic-molecular theory.

11. In terms of molecules, explain what happens when a liquid evaporates.

12 What is meant by *the boiling point of a liquid*?

13 Describe the process of freezing of water in terms of molecular motion.

14 Predict whether the pressure in an automobile tire will increase or decrease as the automobile travels from sea level to a high mountain. Assume that the temperature of the gas inside the tire remains constant. (Remember that the pressure inside a tire, as is usually measured, is really the difference between the pressure of the gas inside and outside the tire.)

15 Fill in the following blanks:

As the temperature rises, molecules move about more (a) _____. If the substance is a solid, the molecules (b) _____ about a fixed position; if the substance is a liquid, the molecules move about more (c) _____ but are still held near each other; if the substance is a gas, the molecules move in (d) _____. In a gas, pressure is produced by the molecules (e) _____ with the walls of the container. When a liquid evaporates, the molecules (f) _____ the surface of the liquid and pass into the gaseous state.

16 The temperature at which the vapor pressure of a liquid is equal to the atmospheric pressure is called the _____ of the liquid.

17 When the temperature of a liquid increases, the vapor pressure of the liquid _____.

18 Why is it dangerous to throw an empty spray can into a fire?

19 Temperature is a consequence of the _____.

20 As the temperature is decreased, molecules _____. Since this is so, it is conceivable that a temperature might exist at which the molecules have _____. (Such a temperature does exist and is called *absolute zero*.)

SOLUTIONS

CHAPTER TOPICS

When you have completed this chapter, you should have mastered

The concept of a *solution*

The terminology that applies to solutions

The concept of *concentration*

The meaning of common concentration units

Problem-solving techniques using percent by weight and molarity

10-1 THE CONCEPT OF A SOLUTION

Everyone is somewhat familiar with the word *solution*. If a small amount of sugar is placed in water, the sugar will dissolve. After a sufficiently long time, or after mechanical mixing, the sugar will become uniformly dispersed throughout the water. This dispersion of one substance in another is called a *solution*. The dispersed material (i.e., the sugar) is called the *solute*. The material in which the solute is dispersed (i.e., the water) is called the *solvent*.

There are many familiar examples of solutions. Brine is a solution of solid sodium chloride dissolved in water. This is another example of a solid dissolved in a liquid. Wine is a complex solution containing principally alcohol dissolved in water. This is an example of one liquid dissolved in another liquid. The aerating equipment in an aquarium is used to ensure that a sufficient amount of air is dissolved in the water. This is an example of a gas dissolved in a liquid. The air which we breathe can also be classified as a solution since it contains principally oxygen gas dispersed in nitrogen gas. Air is an example of a solution composed of gases. Many metal *alloys* (e.g., sterling silver and brass) are examples of solutions composed of one or more solids dispersed in another. The most common types of solutions are usually formed from a solid dissolved in a liquid, a liquid dissolved in another liquid, a gas dissolved in a liquid, a gas dissolved in another gas, or a solid dissolved in another solid. (On the molecular level, the molecules of solute become dispersed among the molecules of solvent as shown in Fig. 10-1.)

A solution is not a chemical compound in the usual sense. It is a molecular mixture of two or more substances dispersed in one another. In most solutions there is no true *chemical reaction* between the solute and the solvent. The composition of a solution can be varied (more solute can be added), and the component substances usually can be separated by physical processes (for example, the water can be evaporated from a sugar-water solution, leaving just the solid sugar).

10-2 CONCENTRATION OF SOLUTIONS

In chemistry it is frequently necessary to express the *concentration* of a solution in a quantitative way. Many people mistakenly use the word "strength" when describing the concentration of a solution. Thus a *"strong"* salt water solution would be described better as a *concentrated* salt water solution.

Concentration may be expressed in many different units. A concentration unit involves a quantity (or weight) unit and a volume unit. The quantity unit

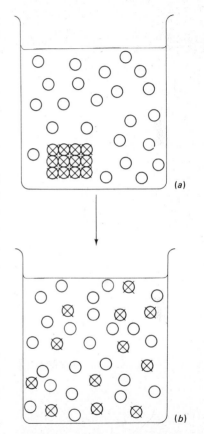

FIGURE 10-1 Solid dissolving in a liquid: (*a*) before solute dissolves; (*b*) after dissolving and mixing

refers to the solute and may be expressed in pounds, grams, moles, or any other appropriate unit. The volume unit usually refers to the volume of the solution formed after the solute has dissolved in the solvent. For example, if 2 lb of sugar are dissolved in enough water to make a total solution volume of 5 gal, then the concentration of sugar in the solution could be expressed as 2 lb divided by 5 gal, or 0.40 lb/gal. This concentration is the same whether we refer to 1 drop or 100 gal of the solution. The *value* of the concentration (0.40 lb/gal) means that each gallon of the solution contains 0.40 lb of dissolved sugar.

A more concentrated solution could be prepared by dissolving 4.0 lb of sugar in enough water to make 5.0 gal of solution. Here the concentration would be 4.0 lb divided by 5.0 gal, or 0.80 lb/gal. In this second solution the sugar molecules are more crowded together than they were in the first solution. The concentration of a solution, then, is a measure of how crowded the solute molecules are. A concentration is usually calculated by using the formula

$$\text{Concentration} = \frac{\text{quantity of solute}}{\text{volume of solution.}}$$

EXAMPLE   Calculate the concentration of sugar, expressed in pounds/gallon, if 2.5 lb of sugar are dissolved in enough water to make 8 gal of solution:

$$\text{Concentration} = \frac{\text{quantity of solute}}{\text{volume of solution}} = \frac{2.5 \text{ lb}}{8.0 \text{ gal}} = 0.312 \text{ lb/gal}$$

Therefore, the concentration is 0.31 lb/gal.

Another type of problem that may be encountered is one in which the concentration of a solution and a certain volume of the solution are known. With this information the amount of dissolved solute may be calculated.

EXAMPLE

Lemonade has a sugar concentration of 1.5 lb/gal. How much sugar is dissolved in a glass which has a volume of 8 fluid oz (or $\frac{1}{16}$ gal)?

The concentration 1.5 lb/gal means that 1 gal of lemonade contains 1.5 lb of dissolved sugar. In the glass, we have only $\frac{1}{16}$ gal of the solution; therefore

$$\text{Amount of dissolved sugar in glass} = (\tfrac{1}{16}\,\cancel{\text{gal}}) \left(\frac{1.5\text{ lb}}{1\,\cancel{\text{gal}}}\right) = 0.094\text{ lb}$$

10-3 MOLARITY AND OTHER CONCENTRATION UNITS

Similar types of calculations may be carried out using any appropriate concentration units. In chemistry, concentration units are usually expressed in metric units, for example, in grams/liter, milligrams/milliliter, or moles/liter.

EXAMPLE If 0.500 liter of seawater is boiled until all the water is removed, the solid residue includes 15.5 g of sodium chloride. Calculate the concentration of sodium chloride in seawater expressed in grams/liter.

$$\text{Concentration of NaCl in seawater} = \frac{15.5 \text{ g}}{0.500 \text{ liter}} = 31.0 \text{ g/liter}$$

In a manner similar to that previously illustrated in the lemonade example, (p. 217) we can calculate the weight of sodium chloride in any given volume of seawater.

EXAMPLE It has been estimated that the total volume of all the oceans of the earth is approximately 1.31×10^{20} liters. If the concentration of sodium chloride in seawater is 31.0 g/liter, calculate the total weight of NaCl contained in all the oceans of the earth:

$$1.31 \times 10^{20} \text{ liters} \times 31.0 \text{ g/liter} = 40.6 \times 10^{20} \text{ g} \quad \text{or} \quad 4.06 \times 10^{21} \text{ g}$$

In chemistry perhaps most frequently used concentration unit is *moles of solute/liter of solution*. This unit is even given a special name, *molarity*, which is sometimes abbreviated as M. Molarity is defined as the number of moles of solute dissolved in one liter of the solution. The molarity unit is quite convenient to use because it is a measure of the number of solute molecules present in one liter of a solution expressed in terms of Avogadro's number of molecules. The molarity of a solution can be calculated from

$$M = \frac{\text{moles of solute}}{\text{volume (in liters) of solution}}$$

Since many chemical reactions occur in solution, the molarity unit provides a convenient way of measuring the number of molecules that will be available for chemical reaction.

EXAMPLE Seawater contains 31 g/liter of NaCl. Calculate the molarity of the NaCl.

Step 1 Calculate the molecular weight of NaCl:

$$\begin{array}{r} 23.0 \text{ awu} \\ 35.5 \text{ awu} \\ \hline 58.5 \text{ awu} \end{array}$$

Therefore, 1 mole of NaCl = 58.5 g.

Step 2 Calculate the number of moles/liter of NaCl:

$$\text{Molarity of NaCl} = (31.0 \text{ g/liter}) \left(\frac{1 \text{ mole}}{58.5 \text{ g}}\right) = 0.53 \text{ mole/liter}$$

EXAMPLE The molarity of glucose in human blood is 4.4×10^{-3} M, and the total blood volume of an average person is about 5 liters. Calculate the total weight of glucose ($C_6H_{12}O_6$) dissolved in the blood of an average person

Step 1 Calculate the molecular weight of glucose:

$C_6H_{12}O_6$

$6 \times 12 = 72$ awu

$12 \times 1 = 12$ awu

$6 \times 16 = \underline{96 \text{ awu}}$

180 awu

Therefore, 1 mole of glucose = 180 g.

Step 2 Calculate the total number of moles of glucose in the total blood volume:

$$4.4 \times 10^{-3} \text{ mole/}\sout{\text{liter}} \times \frac{5 \, \sout{\text{liters}}}{1} = 22.0 \times 10^{-3} \text{ moles}$$

Step 3 Calculate the weight of the glucose:

$$\frac{22.0 \times 10^{-3} \, \sout{\text{moles}}}{1} \times \frac{180 \text{ g}}{\sout{\text{mole}}} = 3{,}960 \times 10^{-3} \text{ g} = 3.96 \text{ g}$$

It is frequently desired to prepare a less concentrated solution from a more concentrated solution. This can be done by *dilution*. Dilution problems can frequently be solved by using the equation

$$M_1 V_1 = M_2 V_2$$

where M_1 = molarity of the original solution
V_1 = volume of the original solution
M_2 = molarity of the final solution
V_2 = volume of the final solution

This is a true equation because we are *not* changing the number of moles of solute present. We are only making the solute particles less crowded by adding more solvent to increase the total volume of the solution.

EXAMPLE

Concentrated hydrochloric acid contains 12 moles of HCl/liter of solution (12 M). Calculate how many milliliters of concentrated HCl must be used to prepare 300 ml of 2.0 M hydrochloric acid. (*Hint*: The total number of moles of HCl in the 300 ml of 2.0 M HCl must be equal to the total number of moles of HCl in the necessary volume of 12 M HCl.)

Step 1 Calculate the number of moles of HCl in the dilute solution:

$$\text{Moles of HCl} = (3.0 \times 10^2 \text{ ml})(1 \text{ liter}/10^3 \text{ ml})(2.0 \text{ moles/liter})$$

$$= 6.0 \times 10^{-1} \text{ mole} = 0.60 \text{ mole}$$

Step 2 Calculate the volume of 12 M HCl to supply 0.60 mole of HCl:

$$0.60 \text{ mole} = (12.0 \text{ moles/liter})(V)$$

$$\text{or } V = \frac{0.60 \text{ mole}}{12.0 \text{ moles/liter}} = 0.050 \text{ liter}$$

Then $(0.050 \text{ liter})(10^3 \text{ ml/liter}) = 50.0$ ml of 12 M HCl is needed.

Alternatively, the problem may be worked by using the equation $M_1V_1 = M_2V_2$ and solving for V_1:

$$V_1 = \frac{M_2V_2}{M_1} = \frac{(2.0 \text{ moles/liter})(0.300 \text{ liter})}{12.0 \text{ moles/liter}} = 0.50 \text{ liter} \quad \text{or } 50.0 \text{ ml}$$

10-4 PERCENT BY WEIGHT

A concept closely related to the concentration units we have been dealing with is the *percent by weight* of a solute in a solution. The conceptual difference between the percent by weight and grams/liter of a solution is that percent by weight refers to the weight of solute in a given *weight* (100 g) of solution while grams/liter refers to the weight of solute in a given *volume* (1 liter) of solution.

In problems involving conversion from grams/liter to percent by weight, an additional bit of information is always needed: the *density* of the solution, which relates the weight and the volume of the solution.

EXAMPLE

Calculate the percent by weight of sucrose in sugar-cane juice if the concentration of sucrose is 159 g of sucrose/liter of juice (159 g/liter) and the density of the solution is 1.06 g/ml.

Step 1 The total weight of 1.0 liter of the sugar-cane juice is

$$(1.06 \text{ g/ml}) (1000 \text{ ml/liter}) = 1{,}060 \text{ g/liter}$$

or 1 liter weighs 1,060 g.

Step 2 Since 159.0 g of sucrose is contained in 1,060 g of solution (note that both of these units refer to 1 liter of the solution), the percent by weight will be

$$\frac{1.59 \times 10^2}{1.060 \times 10^3} \times 10^2 = 1.5 \times 10^1 = 15\%$$
$\uparrow$
(To convert to percentage)

Since 15 percent of the solution is sucrose, the remaining 85 percent of the solution is water.

It is also possible to calculate the molarity of a solution if both the percent by weight of the solute and the density of the solution are known.

EXAMPLE   Vinegar is a solution which contains 5.00% (by weight) hydrogen acetate dissolved in water. The density of vinegar is 1.01 g/ml. Calculate the molarity of hydrogen acetate in vinegar.

Step 1   Calculate the weight of 1.00 liter of solution:

$$\text{Weight} = (1 \text{ liter})(10^3 \text{ ml/liter})(1.01 \text{ g/ml}) = 1.01 \times 10^3 \text{ g}$$

Step 2   Calculate the weight of $H(C_2H_3O_2)$ in 1.00 liter of vinegar:

$$\begin{aligned}\text{Weight of } H(C_2H_3O_2) &= (1.01 \times 10^3 \text{ g})(5\%) \\ &= (1.01 \times 10^3 \text{ g})(0.0500) \\ &= \mathbf{5.05 \times 10^1 \text{ g}}\end{aligned}$$

Step 3   Calculate the moles of $H(C_2H_3O_2)$ contained in 5.05×10^1 g $H(C_2H_3O_2)$:

Molecular weight of $H(C_2H_3O_2)$ = 1.01
24.00
3.03
32.00
―――
60.00 awu

Therefore, 1 mole of $H(C_2H_3O_2)$ = 60.0 g.

Moles of $H(C_2H_3O_2)$ = $(5.05 \times 10^1 \text{ g}) \left(\dfrac{1 \text{ mole}}{60.0 \text{ g}}\right)$ = 0.842 mole

Step 4 Calculate the molarity of the solution:

$$\dfrac{0.842 \text{ mole}}{1 \text{ liter}} = 0.842 \ M$$

See Table 10.1 for a summary of these types of calculations.

TABLE 10-1 Types of calculations presented in this chapter

$C = \dfrac{W}{V}$	The symbol C represents the concentration of the solution. W is the weight of dissolved solute expressed in any desired weight unit (pounds, grams). The volume V may be expressed in any volume units (gallons, liters).

TABLE 10-1 (Continued)

$W = C \times V$ The same statements apply here as in the above calculations. (*Note:* A similar calculation is: moles = $M \times V$, where M is molarity and V is the volume of the solution in liters.)

$$M = \frac{\text{moles of solute}}{\text{volume of solution}}$$

The symbol M represents the molarity of the solution. The units of M are moles/liter when the volume is expressed in liters.

Dilution: $M_1 \times V_1 = M_2 \times V_2$

In this equation, M_1 = starting molarity
V_1 = starting volume
M_2 = final molarity
V_2 = final volume

$$\text{Percent by weight} = \frac{C(\text{g/liter}) \times V(\text{liter}) \times 10^2}{d(\text{g/ml}) \times 10^3(\text{ml/liter}) \times V(\text{liter})}$$

Remember, *C is the concentration of solute in grams/liter*, and *d* is the *density* of the solution in *grams/milliliter*; 10^2 is included to convert the decimal value to percentage.

GLOSSARY

Alloy: A solid solution of two or more metals

Concentration: A quantitative measure of the quantity of solute contained in a given volume of a solution

Dilution: The preparation of a less concentrated solution from a more concentrated solution usually accomplished by adding pure solvent to the more concentrated solution

Molarity: The number of moles of solute dissolved in one liter of a solution (often abbreviated *M* and expressed in units of moles/liter)

Solute: The dispersed material in a solution

Solution: The homogeneous dispersion of one substance in another

Solvent: The dispersing medium of a solution

SELF TEST

1. Give two ways that a solution could be formed from one solid, one liquid, and one gas.

2. The dispersing medium of a solution is called the _____.

3. The substance which is dispersed in a solution is called the _____.

4   The _____ is a measure of how crowded together the solute molecules are in a solution.

5   A brine solution contains 250 g of NaCl in 1 liter of solution. How many grams of NaCl are contained in 300 ml of the brine?

6   The molarity of a solution is defined as _____.

7   A bottle is labeled as 2.0 M NaCl. This means that 1.0 liter of the solution contains (a) _____ mole(s) of NaCl. (b) How many grams of NaCl are in 1.0 liter of the 2.0-M solution?

8   Stomach acid is approximately 0.10 M HCl. How many grams of HCl are contained in 2.0 liters of stomach acid?

9   A volume of 50.0 ml of 12.0 M HCl is diluted to 2.0 liters by adding pure water. What is the final concentration of HCl?

10 What is the percent by weight of a water solution which is 0.80 M in $H(C_2H_3O_2)$? (The density of the solution is 1.01 g/ml.)

ANSWERS

1   Solid in liquid solution; gas in liquid solution
2   Solvent

3   Solute
4   Concentration
5   75 g of NaCl
6   The number of moles of solute in one liter of a solution
7   (a) 2; (b) 117 g of NaCl
8   7.3 g of HCl
9   0.3 mole/liter, or 0.3 M
10 4.75% $H(C_2H_3O_2)$

EXERCISES

1   Define the following terms:
    (a) Solute
    (b) Solution
    (c) Concentration
    (d) Molarity
    (e) Percent by weight

2   Give three different units used to express the concentration of a solution.

3   The concentration of a solution is expressed in pounds/gallon. Using conversion factors from a table of conversion factors, calculate the concentration of the solution expressed in grams/milliliter.

4   If the concentration of cows on a particular ranch is three cows per acre

and the ranch consists of 250 acres, how many cows are on the ranch? (Note the analogy to the concentration and volume of a solution.)

5. If 4.0 g of $H_2(SO_4)$ are dissolved in enough water to make 2.0 liters of solution, what is the concentration of the solution expressed as
   (a) Grams/liter
   (b) Moles/liter

6. If 18 g of $H_2(SO_4)$ are dissolved in enough water to make 3.0 liters of solution, what is the concentration of the solution expressed as
   (a) Grams/liter
   (b) Moles/liter

7. If 0.020 g of magnesium hydroxide (milk of magnesia) is dissolved in 2.0 liters of solution, what is the molarity of the solution?

8. If 0.35 g of calcium hydroxide (quicklime) is dissolved in 3.0 liters of water, what is the molarity of the solution?

9. What volume of 4.0 M $H_2(SO_4)$ is needed to supply 0.0050 mole of pure $H_2(SO_4)$?

10. What volume of 0.1 M H_2SO_4 is needed to supply 0.30 mole of H_2SO_4?

11. What volume of 0.50 M H_2SO_4 is needed to supply 15.0 g of pure H_2SO_4?

12. What would be the molarity of a solution of $H(NO_3)$ if 3.0 g of $H(NO_3)$ is present in 90 ml of solution?

13. A solution of $H_3(PO_4)$ is 0.35 M. Calculate the number of
    (a) Moles of H_3PO_4 in 200 ml of solution
    (b) Grams of H_3PO_4 in 200 ml of solution

14. A solution of H_2SO_4 is 0.40 M. Calculate the number of
    (a) Moles of H_2SO_4 in 100 ml of solution
    (b) Grams of H_2SO_4 in 100 ml of solution

15. An isotonic sucrose solution is 9.0% sucrose by weight. Calculate the weight of sucrose necessary to prepare 2.0 liters of isotonic solution whose density is 1.03 g/ml.

16. An isotonic sucrose solution is 9.0% sucrose by weight. Calculate the weight of sucrose needed to prepare 1.5 liters of isotonic solution whose density is 1.03 g/ml.

17. The molarity of glucose in human blood is about 4.4×10^{-3} M. Calculate the number of
    (a) Moles of glucose in 1.0 liter of blood
    (b) Grams of glucose in 1.0 liter of blood
    (c) Moles of glucose in 1 pt of blood

18. If a man drinks one cocktail containing 14 g of ethyl alcohol (formula: C_2H_5OH), calculate the molarity of alcohol in the bloodstream, assuming all the alcohol is absorbed into the blood and the total blood volume of the man is 5.0 liters.

19. The molarity of glucose in human blood is about $4.4 \times 10^{-3}\,M$. Calculate the number of
    (a) Moles of sucrose contained in a 25.0-ml blood sample
    (b) Grams of glucose in a 25.0-ml sample

20. If the concentration of sugar in a lemonade concentrate is 7.5 lb/gal, calculate the volume of concentrate needed to prepare 1 qt of lemonade with a sugar concentration of 1.5 lb/gal.

21. Calculate the molarity of HCl in a solution which is 20.2% (by weight) HCl and has a density of 1.10 g/ml.

22. What is the final concentration of a solution prepared by diluting 50.0 ml of 12 M HCl to 150.0 ml by adding pure water?

23. Muriatic acid is a commercial grade of hydrogen chloride dissolved in water. The concentration of HCl is about 10.0 M. Calculate the volume of muriatic acid needed to prepare 3.0 liters of 0.5 M HCl.

24. An aqueous solution of sodium nitrate has a density of 1.07 g/ml and is 10 percent by weight.

(a) Calculate the molarity of sodium nitrate.
(b) How many grams of sodium nitrate would be contained in 350 ml of this solution?

25 Pure sulfuric acid (H_2SO_4) has a density of 1.83 g/ml.
(a) How many grams of pure H_2SO_4 are required to prepare 500.0 ml of a 6.00-M solution of H_2SO_4 in water?
(b) How many milliliters of pure sulfuric acid are needed to prepare the solution in part (a)?

26 The percent by weight of salt in the Great Salt Lake is about 25 percent. The density of this solution is 1.20 g/ml. Calculate the
(a) Number of grams of salt contained in 150 ml of this solution
(b) Molarity of this solution

27 A sugar solution contains 93.0 g of sugar in 1 liter. How many milliliters of this solution are required to make 250.0 ml of a sugar solution having a concentration of 31.0 g of sugar in 1 liter?

28 An injection of 0.30 ml of a sedative solution with a concentration of 60 mg/ml was administered to a child having a blood volume of approximately 2.0 liters. Calculate the concentration of the sedative in the blood supply when it is uniformly distributed.

29 The density of the sulfuric acid in an automobile battery that is fully

charged is about 1.30 g/ml. The percent of H_2SO_4 in this solution is 40%. Calculate the
(a) Number of grams of H_2SO_4 in 500 ml of this solution
(b) Molarity of the solution

30. At atmospheric pressure, 0.0039 g of O_2 is dissolved in 100 ml of water at 25°C. (Assume the density of the solution to be 1.00 g/ml.) Calculate the
(a) Weight percent of O_2 in the solution
(b) Molarity of O_2 in the solution

31. The density of an air sample is 1.29×10^{-3} g/ml. If the air sample is about 24% by weight O_2, calculate the
(a) Weight of O_2 in 1.00 liter of air
(b) Molarity of O_2 in air

ACIDS, BASES, AND SALTS

ACIDS, BASES, AND SALTS

CHAPTER TOPICS

When you have completed the study of this chapter, you should

Know what an *acid* is

Know what a *base* is

Know what a *salt* is

Be able to correctly name common acids, bases, and salts

Be able to recognize common inorganic acids, bases, and salts from their chemical formulas

Be able to write chemical equations for neutralization reactions

Be able to write simple ionic equations

Understand the concept of *normality*

Be able to work problems involving normality, volume, and number of equivalents

11-1 ACIDS

Many common compounds are either acids, bases, or salts. These are among the compounds that we most frequently encounter. For example, any food with a sour taste falls into the category of an *acid*. Vinegar is a very common acid with which we are all familiar. Lemons, oranges, and other citrus fruits also contain acids. The human stomach contains hydrochloric acid. The liquid in an ordinary lead storage battery is sulfuric acid. Phosphoric acid is used in the manufacture of phosphate fertilizers. Thus acids are frequently encountered in everyday life.

All of these substances that we call acids have similar properties. The feature common to all water solutions of acids, which causes the similar properties, is the presence of the hydrogen ion. In fact, this is the simplest definition of an acid: *An acid is a substance capable of furnishing hydrogen ions in a water solution.* When hydrogen ions are present in a water solution, one hydrogen ion will tend to associate with *one or more* water molecules. When one hydrogen ion is associated with one water molecule, the resulting species has the formula H_3O^+ and is called the *hydronium ion*. For our purposes, the hydrogen ion in water solution will be represented simply as H^+. We should bear in mind, however, that the H^+ ion, when present in a water solution, is associated with one or more water molecules. (Some properties of acids are summarized briefly in Table 11-1.)

In the chemical formula of an acid the symbol of the replaceable hydrogen is written first. The formulas of some common acids, together with their names,

TABLE 11-1 Properties of acids

Contain H^+ in water solution
Will react with bases to form salts
Will react with active metals to liberate hydrogen gas
Turn litmus (an acid-base indicator) red
Impart a sour taste to foods

TABLE 11-2 Some common acids

FORMULA	NAME OF PURE COMPOUND	NAME AS AN ACID
HCl	Hydrogen chloride	Hydrochloric acid
*HNO_3	Hydrogen nitrate	Nitric acid
*H_2SO_4	Hydrogen sulfate	Sulfuric acid
*H_3PO_4	Hydrogen phosphate	Phosphoric acid
*$HC_2H_3O_2$	Hydrogen acetate	Acetic acid

Note: The polyatomic ions have the parentheses omitted.

TABLE 11-3 Common binary acids

FORMULA	NAME OF PURE COMPOUND	NAME AS AN ACID
HCl	Hydrogen chloride	*Hydro*chlor*ic* acid
HBr	Hydrogen bromide	*Hydro*brom*ic* acid
HI	Hydrogen iodide	*Hydr*iod*ic* acid
H_2S	Hydrogen sulfide	*Hydro*sulfur*ic* acid

are given in Table 11-2. As can be observed from this table, there are *two ways of naming acids*: (1) In naming the *pure compound*, we first write the word *hydrogen* and then the name of the negative ion. For example, H_2SO_4 is named *hydrogen sulfate*. This method of nomenclature follows the pattern that is used in naming salts and other common compounds. It was discussed in Chap. 5 and has been used in this text until now. (2) The compound may also be named as an acid. For a *binary acid* (i.e., an acid containing two elements, hydrogen and one nonmetallic element), the name is formed by using the prefix *hydro*, the root of the name of the nonmetallic element, and the ending *ic*. (Some common binary acids are listed in Table 11-3.)

Some of the acids listed in Table 11-2 include polyatomic negative ions. Note that the ending *ic* is used for the name of the acid when the name of the negative ion ends in *ate*. The formulas and names that have been given here represent only a few of the most commonly encountered acids. These names and formulas should be mastered before proceeding if they are not already familiar to you.

11-2 BASES

Bases are also frequently encountered compounds. In a sense, bases are the chemical opposites of acids. A base may be defined as *a substance capable of furnishing hydroxide ions in an aqueous solution*. The simplest type of base consists of a positive ion associated with an hydroxide ion. Familiar examples of bases include lye, NaOH; slaked lime, $Ca(OH)_2$; and milk of magnesia, $Mg(OH)_2$. These bases all share certain similar properties. These are listed in Table 11-4.

TABLE 11-4 Properties of bases

Contain OH^- in water solution
Will react with acids to form salts
Turn litmus blue
Impart a bitter taste to foods
Have a characteristic "slippery feeling"

TABLE 11-5 Some common bases

FORMULA	NAME	OTHER NAMES
$Mg(OH)_2$	Magnesium hydroxide	Milk of magnesia
$NaOH$	Sodium hydroxide	Lye
KOH	Potassium hydroxide	Caustic potash
$Ca(OH)_2$	Calcium hydroxide	Slaked lime

When writing the chemical formula of a base, the convention has been adopted of writing the symbol of the positive ion first, followed by the hydroxide ion. Table 11-5 shows the formulas and names of some common bases. The name of a base is easily obtained by first writing the name of the positive ion and then adding the word *hydroxide*.

Some bases do not contain hydroxide ions unless they are dissolved in water. Ammonia, a gas in the pure state at room temperature, is such a compound. When ammonia dissolves in water, however, hydroxide ions are formed, as shown by the equation

$$NH_3 + HOH \rightarrow NH_4^+ + OH^-$$

(*Note*: This reaction does not proceed to completion.)

11-3 SALTS

Many of the compounds which we have encountered so far in this text are *salts*.

For example, the most common salt of all is sodium chloride (table salt). In a general sense, a salt is an ionic compound with positive ions other than hydrogen and negative ions other than hydroxide. A salt may be defined as *an ionic compound which can be formed by the reaction of an acid with a base.*

We shall describe a few of the most important properties of salts. First we should recognize that all salts are *solid* substances. This is reasonable when we consider the strong forces of attraction that are present between the positive and negative ions. Since these strong forces of attraction exist, high temperatures are required to provide sufficient energy to break up the crystal lattice and cause the solid salt to melt. Salts also tend to be rather *brittle*, which can be understood in terms of the structure of an ionic solid. When a sufficient force is applied to a solid, the ions may be moved with respect to one another, as shown in Fig. 11-1.

Another familiar property of most salts is that they are *strong electrolytes.* This means that either the water solution of a salt or the molten salt is an excellent conductor of an electric current. This conductivity is due to the fact that ions can move through a liquid readily when a voltage is applied.

11-4 REACTIONS OF ACIDS, BASES, AND SALTS

The most important reaction of acids and bases is the *neutralization reaction.* This may be defined as *the reaction between an acid and a base to form water plus a salt.* This is a special case of an *exchange reaction,* discussed in Chap. 7. In this reaction, the hydrogen ions of the acid combines with the hydroxide ion of the

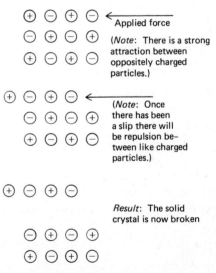

FIGURE 11-1 Fracture of ionic crystal

base to form water. The negative ion from the acid is left with the positive ion from the base, and together they constitute a salt. The following word equation illustrates the neutralization reaction:

Acid plus base yields salt plus water

A more specific example would be

Hydrochloric acid plus sodium hydroxide yields sodium chloride plus water

or

$HCl + NaOH \rightarrow NaCl + HOH$

Additional examples are

Hydrochloric acid plus calcium hydroxide yields calcium chloride plus water

or

$2HCl + Ca(OH)_2 \rightarrow CaCl_2 + 2HOH$

(*Note*: Remember to write correct formulas *first* when writing equations.)

Phosphoric acid plus magnesium hydroxide yields magnesium phosphate plus water

$$2H_3PO_4 + 3Mg(OH)_2 \rightarrow Mg_3(PO_4)_2 + 6HOH$$

All these examples involve *complete neutralization*. In each case there is an implicit assumption that there is a sufficient quantity of acid available to exactly neutralize the base present and vice versa. However, if an acid contains more than one H^+ (that is, if it is a *polyprotic acid*), *partial neutralization* is possible. For example, if 1 mole of sulfuric acid reacts with 1 mole of sodium hydroxide, 1 mole of sodium hydrogen sulfate and 1 mole of water are formed:

$$H_2SO_4 + NaOH \rightarrow NaHSO_4 + HOH$$

If the *acid salt* sodium hydrogen sulfate then contacts additional sodium hydroxide, the second step in the neutralization can occur; thus,

$$NaHSO_4 + NaOH \rightarrow Na_2SO_4 + HOH$$

Complete neutralization of sulfuric acid would be the sum of these two equations:

$$H_2SO_4 + NaOH \rightarrow NaHSO_4 + HOH$$
$$NaHSO_4 + NaOH \rightarrow Na_2SO_4 + HOH$$

$$H_2SO_4 + \cancel{NaHSO_4} + 2NaOH \rightarrow \cancel{NaHSO_4} + Na_2SO_4 + 2HOH$$

Note that NaHSO$_4$ is consumed in the second reaction and is also formed in the first reaction; therefore it *cancels out*. The final equation for complete neutralization is

$$H_2SO_4 + 2NaOH \rightarrow Na_2SO_4 + 2HOH$$

Another characteristic reaction of an acid is the *displacement reaction* in which hydrogen is displaced by an active metal. (This was discussed briefly in Chap. 7.) An example would be

Magnesium plus sulfuric acid yields magnesium sulfate plus hydrogen gas

$$Mg + H_2SO_4 \rightarrow MgSO_4 + H_2$$

Salts will undergo a great many reactions, some of which have been given in Chap. 7. Many salt reactions are exchange reactions which involve the formation of an insoluble substance. An example would be

Silver nitrate plus sodium sulfide yields silver sulfide plus sodium nitrate

$$2AgNO_3 + Na_2S \rightarrow Ag_2S + 2NaNO_3$$

11-5 IONIC EQUATIONS

Because salts are made up of ions, they exist in solution entirely as positive and negative ions. Thus the previous equation can be represented by the *ionic equation*

$$2Ag^+ + 2NO_3^- + 2Na^+ + S^{--} \rightarrow Ag_2S + 2Na^+ + 2NO_3^-$$

It is customary when writing this type of equation to represent insoluble substances, or un-ionized materials, by molecular formulas. The soluble, or ionized, materials are written as separated ions. Since the sodium ions and nitrate ions appear on each side of the arrow, they may be canceled. The result is called a *net ionic equation*:

$$2Ag^+ + S^{--} \rightarrow Ag_2S$$

Note that many of the equations that have been dealt with previously could have been written as either ionic equations or net ionic equations.

A knowledge of the solubilities of various substances allows the chemist to predict which compounds will form insoluble substances. This type of knowledge may be obtained by consulting tables for solubilities, or it may be obtained by experiment. A few elementary solubility rules are summarized in Table 11-6.

TABLE 11-6 Some elementary solubility rules

Most nitrates and acetates are soluble.
Most chlorides are soluble, with the exception of $AgCl$, $PbCl_2$, and Hg_2Cl_2.
Most sulfates are soluble, with the exception of $BaSO_4$, $SrSO_4$, $PbSO_4$, and $CaSO_4$.

TABLE 11-6 Some elementary solubility rules (*Continued*)

Most sulfides are *in*soluble, with the exception of Na_2S, $(NH_4)_2S$, K_2S, MgS, CaS, SrS, and BaS.
Most oxides and hydroxides are *in*soluble, with the exception of NaOH, KOH, NH_4OH; $Ca(OH)_2$, $Ba(OH)_2$, and $Mg(OH)_2$ are only slightly soluble.
Most phosphates and carbonates are *in*soluble, with the exception of the carbonates and phosphates of sodium, potassium, and ammonium.
Most salts of sodium, potassium, and ammonium are soluble.

11-6 NORMALITY

A special unit of concentration that is frequently encountered when dealing with acids and bases is *normality* (N). The *normality of an acid* may be defined as the number of moles of available H^+ per liter of an acid solution. The *normality of a base* may be defined as the number of moles of available OH^- per liter of a basic solution. Thus, if 1 mole of HCl is dissolved in enough water to make 1 liter of solution, the concentration may be expressed as one molar (1 M) or as one normal (1 N). If 1 mole of H_2SO_4 is dissolved in enough water to form 1 liter of solution, the concentration may be expressed as one molar (1 M) or as two normal (2 N) because each mole of H_2SO_4 provides 2 moles of H^+:

$$H_2SO_4 \rightarrow 2H^+ + SO_4^{--}$$

If 1 mole of H_3PO_4 is dissolved in enough water to form 1 liter of solution, the concentration may be expressed as 1 M or as 3 N because each mole of H_3PO_4 provides 3 moles of H^+.

With bases, similar examples may be used: If 0.50 mole of NaOH is dissolved in 1 liter of solution, the concentration may be expressed as 0.50 M or as 0.50 N. If 0.020 mole of $Ca(OH)_2$ is dissolved in 1 liter of solution, the concentration may be expressed as 0.020 M or as 0.040 N because each mole of $Ca(OH)_2$ provides 2 moles of OH^-.

The number of grams of an acid needed to supply 1 mole of H^+ is called *an equivalent*. For HCl the molecular weight is 36.5 awu, and 1 mole (36.5 g) would supply 1 mole of H^+. Therefore, 36.5 g of HCl is *one equivalent* (1 equiv) of HCl. For H_2SO_4 the molecular weight is 98.0 awu; therefore 1 mole (98.0 g) would be 2 equivalent.

The volume of a base with a known normality that is required to neutralize a given volume of an acid with a known normality may be calculated by using the equation

$$V_{acid} \times N_{acid} = V_{base} \times N_{base}$$

In this equation V_{acid} is the volume of the acidic solution, N_{acid} is the normality of the acidic solution, V_{base} is the volume of the basic solution, and N_{base} is the normality of the basic solution. If any three of these quantities are known, the fourth one can be calculated.

EXAMPLE

Calculate the volume of 0.15 N NaOH that is needed to neutralize 100 ml of vinegar which is 0.83 N in $HC_2H_3O_2$.

Step 1 For neutralization, the number of moles of available H^+ must be equal to the number of moles of OH^- added:

$$\text{Moles of available } H^+ = (0.100 \text{ liter})(0.83 \text{ mole of } H^+/\text{liter})$$
$$= 0.083 \text{ mole of } H^+$$

Therefore, we need 0.083 mole of OH^- for neutralization.

Step 2 Calculate the volume of base necessary to supply 0.083 mole of OH^-:

$$0.083 \text{ mole of } OH^- = (V \text{ liters})(0.15 \text{ mole of } OH^-/\text{liter})$$

Therefore V liters $= 0.083$ mole of $OH^-/0.15$ mole of $OH^-/\text{liter} = 0.552$ liter $= 552$ ml.

This type of calculation may also be carried out using the equation

$$V_{acid} \times N_{acid} = V_{base} \times N_{base}$$

Solving for V_{base},

$$V_{base} = \frac{V_{acid} \times N_{acid}}{N_{base}}$$

Substituting the known quantities,

$$V_{base} = \frac{(100 \text{ ml})(0.83 \cancel{N})}{0.15 \cancel{N}}$$

$$= 552 \text{ ml}$$

The equation $V_{acid} \times N_{acid} = V_{base} \times N_{base}$ is true because

$$\frac{\cancel{\text{liters}}}{1} \times \frac{\text{equivalents}}{\cancel{\text{liter}}}$$

represents the number of equivalents of acid present in the solution. For exact neutralization this must be equal to the number of equivalents of base available to react or liters/1 × equivalents/liter of base.

A great many acid-base neutralizations can be represented by the net ionic equation

$$H^+ + OH^- \rightarrow HOH$$

Here we can see that the number of moles of H^+ (equivalents of acid) must be equal to the number of moles of OH^- (equivalents of base).

GLOSSARY

Acid: A substance capable of furnishing hydronium ions in a water solution

Acid salt: A partially neutralized acid

Base: A substance capable of furnishing hydroxide ions in an aqueous solution

Binary compound: A compound containing only two elements

Equivalent of an acid: The number of grams of pure acid required to supply one mole of H^+

Equivalent of a base: The number of grams of pure base required to supply one mole of OH^- (*Note*: this will react with one mole of H^+)

Hydronium ion: The ion responsible for acidic properties of aqueous solutions, which can be represented by H_3O^+ or by H^+

Ionic equation: A chemical equation showing substances as ions that exist as ions

Litmus: A substance which turns red in an acidic solution and blue in a basic solution

Net ionic equation: A chemical equation showing only the ions that react

Neutralization reaction: The reaction between an acid and a base which forms water plus a salt

Normality of an acid: The number of moles of available H⁺ per liter of an acidic solution

Normality of a base: The number of moles of available OH⁻ per liter of a basic solution

Polyprotic acid: An acid containing more than one acidic hydrogen per molecule.

Salt: An ionic compound which can be formed by the reaction of an acid with a base

SELF TEST

1 An acid has a (a) _____ taste, contains the element (b) _____ , and turns litmus (c) _____ .

2 A base feels (a) _____ , usually contains the (b) _____ ion, and turns litmus (c) _____ .

3 Classify each of the following compounds as either an acid, a base, or a salt:
   (a) NaOH   (d) $MgCl_2$
   (b) Na_2SO_4 (e) $HC_2H_3O_2$
   (c) HNO_3

4 The chemical reaction between an acid and a base is usually called a (a) _____ reaction. The products formed usually include a(n) (b) _____ and (c) _____.

5 Fill in the missing product or reactant:
(a) 2NaOH + _____ → 2H$_2$O + Na$_2$SO$_4$
(b) HNO$_3$ + KOH → H$_2$O + _____

6 Write ionic equations for the following:
(a) HCl + NaOH → H$_2$O + NaCl
(b) Pb(NO$_3$)$_2$ + Na$_2$SO$_4$ → PbSO$_4$ ↓ + 2NaNO$_3$
(c) FeCl$_3$ + 3KOH → Fe(OH)$_3$ ↓ + 3KCl
(d) NH$_4$Cl + AgNO$_3$ → AgCl ↓ + NH$_4$NO$_3$
(e) H$_2$SO$_4$ + 2KOH → 2H$_2$O + K$_2$SO$_4$

7 Name the following acids:
(a) HBr
(b) H$_2$SO$_4$
(c) H$_3$PO$_4$
(d) H$_2$S
(e) HNO$_3$

8 How many milliliters of 0.12 N H$_2$SO$_4$ are required to neutralize 25.0 ml of 0.15 N KOH?

ANSWERS

ACIDS, BASES, AND SALTS

1   (a) Sour, (b) hydrogen, (c) red

2   (a) Slippery, (b) $(OH)^-$, (c) blue

3   (a) Base, (b) salt, (c) acid, (d) salt, (e) acid

4   (a) Neutralization, (b) salt, (c) water

5   (a) H_2SO_4, (b) KNO_3

6   (a) $H^+ + Cl^- + Na^+ + OH^- \rightarrow H_2O + Na^+ + Cl^-$
    (b) $Pb^{++} + 2NO_3^- + 2Na^+ + SO_4^{--} \rightarrow PbSO_4 \downarrow + 2NA^+ + 2NO_3^-$
    (c) $Fe^{3+} + 3Cl^- + 3K^+ + 3OH^- \rightarrow Fe(OH)_3 \downarrow + 3K^+ + 3Cl^-$
    (d) $NH_4^+ + Cl^- + Ag^+ + NO_3^- \rightarrow NH_4^+ + AgCl \downarrow + NO_3^-$
    (e) $2H^+ + SO_4^{--} + 2K^+ + 2OH^- \rightarrow 2H_2O + 2K^+ + SO_4^{--}$

7   (a) Hydrobromic acid
    (b) Sulfuric acid
    (c) Phosphoric acid
    (d) Hydrosulfuric acid
    (e) Nitric acid

8   31.2 ml

EXERCISES

1. What are the characteristics by which an acid may be recognized?

2. What are the characteristics by which a base may be recognized?

3. What ion is responsible for the properties of a base in water solution?

4. What ion is responsible for the properties of an acid in water solution?

5. Why do salts tend to melt at high temperatures?

6. Why do salts tend to break rather than bend when struck with a hammer?

7. Give the name of the *pure compound* and the name as an acid for each of the following.
   (a) HCl
   (b) HNO_3
   (c) HBr
   (d) H_3PO_4
   (e) $HC_2H_3O_2$
   (f) H_2SO_4

8. Write the formulas of the following acids:
   (a) Sulfuric acid
   (b) Acetic acid

(c) Phosphoric acid
(d) Hydrobromic acid

9 Lime juice has a sour taste. From this fact, what ion must be present in this water solution?

10 Write the formulas of the following bases:
(a) Sodium hydroxide
(b) Lithium hydroxide
(c) Calcium hydroxide
(d) Magnesium hydroxide
(e) Ammonia

11 Classify each of the following compounds as either an acid, a base, or a salt:
(a) Na_2SO_4
(b) $Mg(OH)_2$
(c) $HC_2H_3O_2$
(d) $CaCl_2$
(e) $H_2C_2O_4$

12 Write complete balanced equations for the following reactions, assuming complete neutralization in each case:
(a) Hydrochloric acid reacts with sodium hydroxide.
(b) Sulfuric acid reacts with sodium hydroxide.

(c) Phosphoric acid reacts with lithium hydroxide.
(d) Acetic acid reacts with calcium hydroxide.
(e) Nitric acid reacts with calcium hydroxide.
(f) Nitric acid reacts with aluminum hydroxide.
(g) Phosphoric acid reacts with calcium hydroxide.
(h) Phosphoric acid reacts with aluminum hydroxide.

13 Write balanced equations for the first step in the neutralization of
 (a) Sulfuric acid by sodium hydroxide
 (b) Phosphoric acid by sodium hydroxide
 (c) Calcium hydroxide by hydrochloric acid

14 Calculate the weight of water that can be formed by allowing 5 g of sulfuric acid to react with an excess of sodium hydroxide.

15 Calculate the weight of water that can be formed by allowing 5 g of sodium hydroxide to react with an excess of sulfuric acid.

16 Given that a milk of magnesia tablet contains 0.30 g of $Mg(OH)_2$, calculate the weight of stomach acid (HCl) which may be neutralized by one tablet, assuming all the hydroxide ion is neutralized.

17 Some commercial antacid tablets contain $Al(OH)_3$.
 (a) Write the balanced chemical equation between $Al(OH)_3$ and HCl.
 (b) Explain why $Al(OH)_3$ behaves as an antacid.

(c) How many moles of pure HCl can be neutralized by a tablet containing 0.10 g of $Al(OH)_3$?

(d) How many milliliters of 0.10 N HCl can be completely neutralized by 0.30 g of $Al(OH)_3$?

18 Some commercial antacid tablets contain $MgCO_3$.
(a) Write the balanced chemical equation between $MgCO_3$ and HCl. (*Hint*: See Sec. 7-4.)
(b) Explain why $MgCO_3$ behaves as an antacid.
(c) Why might you burp after taking this antacid?

19 A solution of H_3PO_4 in water is 4.0 M; therefore the solution is _____ N.

20 A solution of H_2SO_4 in water is 3.0 M; therefore the solution is _____ N.

21 A solution of 6.0-N H_2SO_4 is _____ M.

22 Write balanced equations for the following reactions:
(a) $Ba(NO_3)_2 + Na_2SO_4$ (barium sulfate is insoluble)
(b) $ZnCl_2 + (NH_4)_2S$ (zinc sulfide is insoluble)
(c) $AgNO_3 + Na_2CO_3$ (silver carbonate is insoluble)
(d) $Na_3PO_4 + Ca(NO_3)_2$ (calcium phosphate is insoluble)
(e) $Pb(NO_3)_2 + NaCl$ [lead(II) chloride is insoluble]

23 Rewrite the equations in Exercise 22 as *ionic equations*.

24 Rewrite the equations in Exercise 22 as *net ionic equations*.

25 Write a chemical equation showing the separation of ions which occurs when each of the following salts is dissolved in water (the first equation has been completed):
(a) $Na_2SO_4 \rightarrow 2Na^+ + SO_4^{--}$
(b) $KNO_3 \rightarrow$
(c) $CaCl_2 \rightarrow$
(d) $Al(NO_3)_3 \rightarrow$
(e) $Fe_2(SO_4)_3 \rightarrow$
(f) $H_2SO_4 \rightarrow$

26 Fill in the formula of the missing reactant or product:
(a) $Ag^+ + NO_3^- + Na^+ +$ _____ $\rightarrow AgCl + Na^+ + NO_3^-$
(b) $Pb^{++} + 2NO_3^- + 2Na^+ + SO_4^{--} \rightarrow$ _____ $+ 2Na^+ +$ _____
(c) $H^+ + OH^- \rightarrow$ _____
(d) $2Al^{3+} + 6Cl^- + 6Na^+ + 3CO_3^{--} \rightarrow$ _____ $+ 6Na^+ + 6Cl^-$
(e) $2NH_4^+ +$ _____ $+ 2Ag^+ + 2NO_3^- \rightarrow Ag_2S + 2NH_4^+ + 2NO_3^-$

27 With the aid of the solubility rules given in this chapter, write the correct chemical formula of the precipitate which forms when

(a) A solution of sodium chloride is mixed with a solution of lead(II) nitrate
(b) A solution of ammonium sulfide is mixed with a solution of cadmium nitrate
(c) A solution of potassium carbonate is mixed with a solution of calcium chloride

28 Write the net ionic equation for the reaction of
(a) Sodium chloride and lead(II) nitrate
(b) Ammonium sulfide and cadmium nitrate
(c) Potassium carbonate and calcium chloride

29 A volume of 15.0 ml of 0.10 N H_2SO_4 will provide _____ mole(s) of available H^+.

30 A volume of 15.0 ml of 0.10 M H_2SO_4 will provide _____ mole(s) of available H^+.

31 A sample of 5.0 g of $Ca(OH)_2$ will provide _____ mole(s) of available OH^-.

32 A sample of 10.0 g of $Al(OH)_3$ will provide _____ mole(s) of available OH^-.

33. Calculate the normality of a solution containing 18.25 g of HCl per liter of solution.

34. Calculate the normality of a solution containing 196 g of sulfuric acid dissolved in 500 ml of solution.

35. What volume of 0.2 N sodium hydroxide will be needed to exactly neutralize 50 ml of 0.15 N sulfuric acid?

36. What volume of 0.3 N calcium hydroxide will be needed to exactly neutralize 40 ml of 0.2 N phosphoric acid?

37. What volume of concentrated sulfuric acid (approximately 36 N) will be needed to prepare 3 liters of a 2 N solution of sulfuric acid?

38. What volume of concentrated ammonium hydroxide (approximately 16 N) will be needed to prepare 400 ml of a 6 N solution of ammonium hydroxide?

39. A 5.00-ml sample of lemon juice is just neutralized by 25.0 ml of 0.15-N NaOH solution.
    (a) What is the normality of lemon juice?
    (b) If the acid in the lemon juice is principally citric acid ($H_3C_6H_5O_7$), what is the molarity of citric acid in the juice?

EQUILIBRIUM

12 EQUILIBRIUM

CHAPTER TOPICS

Many processes in nature proceed in two directions *at the same time*, and we shall study this kind of process in this chapter. As part of your study you should learn

What is meant by a *dynamic equilibrium*

To recognize the equilibrium nature of some familiar processes

How a system in a dynamic equilibrium may have the position of equilibrium shifted

The meaning of the *strength* of an acid or a base

To recognize the meaning of the *double arrow* in an equation

To understand the concept represented by pH

12-1 DYNAMIC EQUILIBRIUM

A *dynamic equilibrium* can exist whenever two processes are occurring simultaneously if one of these processes is the exact reverse of the other and the two processes are occurring at the same rate. Many familiar processes involve a dynamic equilibrium.

For example, the evaporation of a liquid in a closed container will establish a dynamic equilibrium. When a softdrink is placed in a capped bottle, molecules of water are continually leaving the surface of the liquid and escaping into the gas phase above the liquid. At the same time, molecules of water from the gas phase are continually striking the surface and rejoining the molecules in the liquid. These processes are the exact reverse of each other. When the rate at which the molecules leave the liquid is equal to the rate at which molecules return to the liquid, a dynamic equilibrium has been established (see Fig. 12-1).

Initially, the rate at which molecules leave a liquid will be faster than the rate at which molecules return to the liquid because few molecules are in the gaseous state. As the number of molecules in the gaseous state increases, the rate at which the molecules return to the liquid increases; eventually, the two processes will occur at the same rate.

Similarly, a dynamic equilibrium may be established when a substance melts. An example might be a mixture of ice and water in a perfect Thermos bottle, which permits no heat to be absorbed from the surroundings. When the ice and water are first placed in the Thermos, some of the ice will melt until the water is cooled to 0°C. There will be no net change in the amount of ice present after 0°C is reached, but some ice will be melting and some water will be freezing. This mixture of ice and water at 0°C (with no heat being absorbed from the surroundings) is a dynamic equilibrium mixture. Some molecules will be leaving the surface of the ice, going into liquid; and some molecules will be leaving the liquid, becoming ice. The rate at which molecules are leaving the surface of the ice is equal to the rate at which molecules of the liquid are returning to ice.

When a solid dissolves in a liquid, a dynamic equilibrium is established when the rate at which molecules leave the solid is equal to the rate at which the molecules return to the solid. Under these conditions it appears that no more material is dissolving; such a solution is said to be *saturated*. When a solid dis-

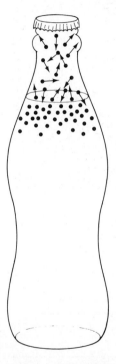

FIGURE 12-1 Equilibrium in a softdrink bottle

solves in a liquid, the process could be as represented by Fig. 12-2. Initially, the rate at which molecules leave the solid is greater than the rate at which molecules return to the solid; therefore, the solid dissolves faster initially and dissolves more slowly as equilibrium is approached (as may be observed if an excess of sugar is placed in a glass of iced tea).

Equilibrium may be established in a similar manner when one chemical reaction is the exact reverse of another. One example of a process that may establish a reversible equilibrium is the case in which carbon dioxide reacts with water:

$$H_2O + CO_2 \rightleftharpoons H_2CO_3$$

When the H_2O and CO_2 first come into contact, the rate of the forward reaction (from left to right) is greater than the reverse reaction (from right to left). As the amount of H_2CO_3 present increases, the rate of the reverse reaction increases. When the rate at which the water combines with carbon dioxide is equal to the rate at which the carbonic acid decomposes, a dynamic equilibrium exists.

Be sure to note the double arrow ($\rightleftharpoons$) in the equation. This symbol is used to indicate that the forward reaction is occurring *and* that the reverse reaction is *also* occurring at the same time. In such a system a dynamic equilibrium may be established if the rates of the two reactions become equal. A reaction that can go in either direction is said to be a *reversible reaction*.

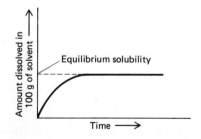

FIGURE 12-2 Rate at which a solid dissolves in a liquid

12-2 SHIFTING THE POSITION OF EQUILIBRIUM

Since the types of processes we have been considering are *dynamic* (i.e., the forward reaction is still occurring and the reverse reaction is still occurring), it is not too surprising to find that the position of an equilibrium *may be shifted*. This shift occurs when some factor *changes the rate* of one reaction more than it changes the rate of the opposite reaction.

For example, let us return to the example of the ice-water mixture in a Thermos bottle. If heat were introduced into the system, the rate at which the ice melts would increase, which would shift the position of the equilibrium toward liquid water. However, if enough ice is available, a new equilibrium might be established. At the new equilibrium condition, there will be less ice and more water than there was at the original equilibrium.

This concept, which involves the possibility of a dynamic equilibrium being shifted, is called the *law of Le Châtelier* after Henry Louis Le Châtelier, who first stated it clearly. This law states: "If a system that is in a dynamic equilibrium is subjected to some stress, the equilibrium will shift in the direction that relieves the stress."

At this point it would be interesting for the student to reconsider each example of a dynamic equilibrium that has been discussed. See if you can think of a type of stress that might shift the position of the equilibrium, and think through the effect of that stress. (In general it should be realized that any factor that increases the rate of the forward reaction more than the rate of the reverse reaction will result in an increase in the amount of product at the expense of the reactants.)

12-3 STRENGTHS OF ACIDS AND BASES

The average person is likely to have an incorrect concept of the chemist's mean-

TABLE 12-1 Some strong and weak acids

STRONG ACIDS		WEAK ACIDS	
HCl	Hydrochloric acid	HF	Hydrofluoric acid
HNO_3	Nitric acid	H_2S	Hydrosulfuric acid
H_2SO_4	Sulfuric acid	$HC_2H_3O_2$	Acetic acid
		$H_3C_6H_5O_7$	Citric acid
		H_3PO_4	Phosphoric acid

ing of a *strong acid*. Most people think of a strong acid as being a very corrosive substance. To the chemist, however, a strong acid is one that completely separates into ions when dissolved in water. (It is true that acids which are completely ionized in a water solution are corrosive substances.) Some acids that are extremely corrosive are *not* completely ionized in a water solution, however, and so the chemist would call these *weak acids*. For example, the compound HF does not completely ionize when dissolved in water, and so the chemist regards this as a weak acid even though it is very corrosive. In Table 12-1 common acids are classified as either strong or weak according to the extent to which they ionize in water.

When a weak acid is dissolved in water, a reversible reaction is quickly established. For example, the equilibrium

$$HC_2H_3O_2 \rightleftharpoons H^+ + C_2H_3O_2^-$$

exists in vinegar, which is a solution of acetic acid dissolved in water. If 0.830

mole of pure acetic acid is dissolved in enough water to make 1.00 liter of solution, the equilibrium is established almost immediately. It is found by experiment that the equilibrium concentration of un-ionized acetic acid molecules is 0.827 mole/liter. The concentration of the H^+ and the acetate ion is 0.003 mole/liter. Note that one hydrogen acetate molecule separates to form one H^+ and *also* one $C_2H_3O_2^-$. At equilibrium the rate at which the hydrogen acetate molecules are separating into ions is exactly the same as the rate at which the ions are recombining to form molecules. The total amount of ionized hydrogen acetate *plus* the un-ionized hydrogen acetate equals the quantity originally introduced.

If, now, 1.00 mole of $NaC_2H_3O_2$ (which is completely ionized) is added to the equilibrium system described above, it will constitute a *stress* on the system and will cause the position of equilibrium to shift to the left (illustrating the law of Le Châtelier). When a new equilibrium is established, there will be approximately 1.5×10^{-5} mole/liter of hydrogen ion, 0.83 mole/liter of un-ionized acetic acid and 1.0 mole/liter of acetate ion. We observe that the concentration of hydrogen ion has decreased from 3×10^{-3} to 1.5×10^{-5} mole/liter and that the concentration of un-ionized acetic acid has increased from 0.827 mole/liter to 0.830 mole/liter. This is what is meant by the statement "The position of equilibrium has been shifted to the left."

(At this time the student might reasonably be expected to make qualitative predictions as to whether concentrations increase or decrease, but calculations leading to the numbers presented in the preceding paragraph are reserved for a general chemistry course.)

Bases may also be classified as *strong bases* (in the sense that the chemist uses the term) if they are completely ionized in a water solution. *Weak bases* are

TABLE 12-2 Some strong and weak bases

STRONG BASES	WEAK BASES
NaOH	NH_4OH
KOH	
LiOH	
$Ca(OH)_2$	
$Mg(OH)_2$	

not completely ionized in a water solution. Table 12-2 lists some common bases classified as strong or weak according to the extent to which they ionize in water.

In general, that part of the hydroxide of any metal which dissolves will be completely ionized in a water solution. The hydroxides of the alkali metals are very soluble in water; therefore their solutions may contain very high concentrations of hydroxide ions. Hydroxides of other metals, such as calcium and magnesium, are less soluble [for example, $Ca(OH)_2$ has a solubility of 1.85 g/liter]. Thus, even though that part of the calcium hydroxide which dissolves is completely ionized, the concentration of OH^- in the solution never becomes very high. Most other metallic hydroxides are quite insoluble; therefore they cannot be used to prepare solutions when high concentrations of OH^- are needed. Those solutions that contain only small concentrations of OH^- are sometimes (rather loosely) thought of as weak bases.

A solution of ammonia in water represents a somewhat different situation. When ammonia dissolves in water, the following reaction occurs:

$$NH_3 + HOH \rightleftharpoons NH_4^+ + OH^-$$

In a 1-M solution about 99.6% of ammonia is present as NH_3 molecules. The other 0.4% exists as NH_4^+. For every NH_4^+ ion formed, one OH^- ion is formed; therefore the OH^- concentration is only about 0.4% as great as the total concentration of ammonia in the solution. Thus ammonium hydroxide is a weak base since only a small fraction of it is ionized. In comparison, if 1 mole of NaOH is dissolved in 1 liter of solution, the concentration of the OH^- in the solution is 1 mole/liter. Thus, sodium hydroxide is a strong base.

12-4   pH: THE ACIDITY OR BASICITY OF A SOLUTION

The term *pH* is frequently encountered when dealing with acids and bases. It is used to indicate whether a solution is acidic, basic, or neutral. pH is best defined by the mathematical expression

$$pH = -\log [H^+]$$

where $[H^+]$ means the concentration of hydrogen ion in moles/liter.

Pure water has a pH of 7 and is considered to be *neutral*. Here the $[H^+] = 1 \times 10^{-7}$ mole/liter, and so the log of $[H^+] = -7.0$, or pH = 7.0. Any acidic material has a pH less than 7; and the lower the pH, the more acidic the solution. Any basic material has a pH greater than 7; and the higher the pH, the more basic the solution. (Table 12-3 gives the pH values of some common solu-

TABLE 12-3   pH values of some common solutions

SOLUTION	pH VALUE
1.0-M NaOH solution	14.0
Household ammonia solution	11.9
Borax solution	9.2
Human blood	7.4
Pure water	7.0
Black coffee	5.0
Wine	3.7
Vinegar	2.4
Stomach acid	1.3
0.10-M HCl solution	1.0

tions.) In a 0.10-M solution of HCl in water, the $[H^+] = 0.10$. The log of 0.10 is -1. Therefore,

$$pH = -\log(10^{-1}) = -(-1) = +1$$
$$= 1$$

(More complete discussions of pH are given in more advanced chemistry courses.)

GLOSSARY

Dynamic equilibrium: A state that exists when two opposing reactions occur simultaneously at the same rate

Law of Le Châtelier: See page 265

pH: -log [H$^+$], used to indicate acidity or basicity of a solution

Reversible reaction: A reaction that can go in either direction

Strong acid: An acid that is completely ionized when dissolved in water

Strong base: A base that is completely ionized when dissolved in water

Weak acid: An acid that is only partially ionized when dissolved in water

Weak base: A base that is only partially ionized when dissolved in water

SELF TEST

1. In a dynamic equilibrium both the _____ and the _____ reaction are taking place at the same rate.

2. The position of a dynamic equilibrium may be shifted by the application of _____.

3. If water and ice are in equilibrium with each other, a stress may be applied and the equilibrium may be shifted by adding _____ to the system.

4   If heat is added to an ice-water equilibrium, it will cause _____.

5   A strong acid, as a chemist uses the term, means an acid that is _____ ionized in a water solution.

6   Acetic acid would be classed as a _____ acid.

7   In the equation $HC_2H_3O_2 \rightleftharpoons H^+ + C_2H_3O_2^-$, the double arrow means that both the forward and reverse reactions may _____.

8   Sodium hydroxide is considered a strong base because it is completely _____.

9   If the pH of a solution is 7.0, the solution is considered to be _____.

10 If the pH of a solution is less than 7.0, the solution is considered to be _____.

ANSWERS
1 Forward, reverse
2 Stress
3 Heat

4   Some of the ice to melt
5   Completely
6   Weak
7   Occur
8   Ionized in a water solution
9   Neutral
10 Acidic

EXERCISES

1   (a) Describe the action of the molecules in the dynamic equilibrium present in a partially full container of gasoline.
    (b) Why must the cap on the container be tightly closed?

2   What is necessary for a dynamic equilibrium to be established?

3   When the cap is removed from a carbonated softdrink and it is allowed to stand, why does the taste go "flat"? (Explain in terms of the law of Le Châtelier.)

4   If a broken, jagged crystal of salt is placed in contact with a saturated solution of salt for a long period of time, the weight of the crystal does not change but the shape becomes very symmetrical. Explain why this occurs in terms of molecular motion.

5   Why does sugar dissolve more quickly when *first* placed in contact with pure water?

6   Describe a saturated sodium chloride solution in contact with solid sodium chloride in terms of a dynamic equilibrium.

7   Since an open bottle of household NH_3 imparts a distinct odor to the air in its vicinity, what must be happening to this reversible reaction

$$NH_3 + H_2O \rightleftharpoons NH_4^+ + OH^-$$

8   How may the position of an equilibrium be shifted?

9   What is a strong acid, as the chemist uses the term?

10 Hydrofluoric acid is a very corrosive acid, but the chemist refers to it as a weak acid. Why?

11 (a) Which solution would be a better conductor of electricity: 0.1 *M* HCl or 0.1 *M* $HC_2H_3O_2$?
    (b) Why?

12 Classify each of the following as strong or weak acids:
    (a) H_2SO_4 (c) $H_3C_6H_5O_6$
    (b) HNO_3 (d) H_3PO_4

13   That portion of magnesium hydroxide which dissolves in water exists as ions; the solubility is 9.0×10^{-3} g/liter or 9.0 mg/liter. What is the concentration of OH^- expressed as moles/liter in a saturated solution of $Mg(OH)_2$? Note that in this solution the calculated concentration of OH^- is quite low although $Mg(OH)_2$ is a strong base.

14   If additional hydroxide ion is added to the saturated solution of magnesium hydroxide in Exercise 13, the concentration of the magnesium ion will _____ and the quantity of undissolved $Mg(OH)_2$ will _____.

15   Write the formula of an acid which is capable of furnishing a high concentration of H^+ ions in a solution.

16   Write the formula of a base which is capable of furnishing a high concentration of OH^- ions in a solution.

17   (a) Write an equation for the reversible equilibrium that exists when the weak acid HCN is dissolved in water.
     (b) How might the position of this equilibrium be shifted?

18   If 1 mole of HCN is dissolved in enough water to make 1 liter of solution, and if the HCN is found to be 0.003 percent ionized, calculate the H^+ ion concentration in moles/liter.

19 Ascorbic acid dissolved in water has a solubility of 333.0 g of ascorbic acid per liter of solution. The formula for ascorbic acid is $HC_6H_7O_6$.
   (a) Calculate the *total* concentration of ascorbic acid in moles/liter in a saturated solution.
   (b) Given the equation for ionization

   $$HC_6H_7O_6 \rightleftharpoons H^+ + C_6H_7O_6^-$$

   and the fact that 1.1% of the ascorbic acid is ionized at equilibrium, calculate the $[H^+]$ in the solution.

20 If 10 g of ascorbic acid are dissolved in 1 liter of solution, this equilibrium will be established:

   $$HC_6H_7O_6 \rightleftharpoons H^+ + C_6H_7O_6^-$$

   If sodium ascorbate is then added to the solution, the concentration of the $C_6H_7O_6^-$ ion will be increased. The H^+ concentration will be _____, and the $HC_6H_7O_6$ concentration will be _____.

21 If a saturated solution of sodium bicarbonate in water is prepared, this equilibrium will be established:

   $$NaHCO_3 \rightleftharpoons Na^+ + HCO_3^-$$

If NaCl is then added, the position of equilibrium will shift to the _____, the concentration of bicarbonate ion in solution will _____, and the amount of solid $NaHCO_3$ will _____.

22 When an excess of solid lime (or calcium hydroxide) is placed in water, a dynamic equilibrium is established:

$$Ca(OH)_2 \rightleftharpoons Ca^{++} + 2OH^-$$

If sodium hydroxide is added to a saturated solution of $Ca(OH)_2$, the concentration of OH^- will _____, the concentration of Ca^{++} will _____, and the amount of solid $Ca(OH)_2$ present will _____.

23 When ammonia gas is dissolved in water, the equilibrium

$$NH_3 + H_2O \rightleftharpoons NH_4^+ + OH^-$$

is quickly established. Addition of NH_3 will cause the NH_4^+ concentration to _____ and the OH^- concentration to _____. Addition of NH_4^+ to the original solution will cause the NH_4^+ concentration to _____, the OH^- concentration to _____, and the NH_3 concentration to _____.

24 Give a mathematical definition of pH.

25 Pure water ionizes to a very slight extent. The equation for the ionization of water is

$$H_2O \rightleftharpoons H^+ + OH^-$$

Only 1.8×10^{-7}% of the water molecules are ionized. There are 55.5 moles of water in 1 liter of pure water. Calculate the
(a) Concentration of the hydrogen ion in moles/liter
(b) pH of water

26 Calculate the pH of a 0.010-M solution of HCl.

27 Calculate the pH of a 0.001-M solution of HNO_3.

28 If acetic acid is dissolved in water, will the pH of the solution increase or decrease if sodium acetate is added?

CALCULATIONS DEALING WITH GASES

CALCULATIONS DEALING WITH GASES

CHAPTER TOPICS

In previous chapters we have discussed the qualitative nature of gases and the kinetic-molecular theory. We shall now deal with quantitative aspects of the behavior of gases. When you have completed this chapter you should know

What a barometer is

How gas pressures are expressed

What is meant by *absolute zero*

What the Kelvin temperature scale is

The mathematical expression for the *general gas law*

How to work problems using the general gas law

What the *universal gas constant* is

What is meant by STP

The mathematical expression for Boyle's law, for Charles' law, and for the law of Gay-Lussac

How to work problems using Boyle's law and Charles' law

What the molar volume of a gas is

How to work problems calculating the volume of gas generated by a chemical reaction

13-1 INTRODUCTION

In Chap. 9 we discussed briefly the nature of the solid, liquid, and gaseous states of matter; of these three states of matter, the gaseous state can be most readily described in mathematical terms. The study of the gaseous state of matter and the mathematical laws that describe the behavior of gases was a very important part of the early development of the atomic theory of matter. The study of gases was also very important in the development of the kinetic-molecular theory. To further our study of gases we shall discuss first the mathematical laws that describe their behavior.

13-2 THE GENERAL GAS LAW

The *general gas law* is a mathematical statement that describes the physical behavior of gaseous substances. This law has been developed as the result of numerous studies carried out on many different gases. One statement of this law is

$$PV = nRT$$

where P = pressure
V = volume
n = number of moles of gas
R = a number called the *universal gas constant*
T = temperature on the Kelvin scale

These terms require additional explanation. The volume V of a gas may be expressed in any volume unit. The unit most frequently used is *liters*. You will recall that 1 liter is approximately equal to 1 qt.

The pressure *P* of a gas is usually measured in millimeters of mercury (mm) or in standard atmospheres (atm). To measure the pressure of a gas, a column of mercury is frequently used (see Fig. 13-1). This arrangement is known as a *barometer*. The height of the column of mercury is a measure of the pressure exerted by the gas pressing down on the mercury outside the tube. The higher the pressure, the higher the column of mercury will be. The evacuated space above the column of mercury is easily produced by filling the glass tube with mercury and inverting it. In effect, there is a vacuum (zero pressure) above the mercury column, and some other pressure on the mercury outside the tube. The difference in pressure supports the column of mercury, and so height of the column is a measure of the difference in pressure. For convenience, a *standard atmosphere* is defined as pressure sufficient to support a column of mercury 760 mm in height. (This is approximately the average pressure exerted by the earth's atmosphere at sea level.)

The number of moles of gas *n*, as we have already discussed (Chap. 6), is really a measure of the number of molecules of gas in the sample being considered since 1 mole represents a fixed number of molecules.

The temperature *T* that is used in this calculation is measured on a temperature scale that is different from the one we ordinarily use. This temperature scale is called the *absolute temperature scale*, or the *Kelvin scale*. To help us understand what this means, we should realize that there is a temperature so low that nothing can be colder. This lowest possible temperature is called *absolute zero*, and the absolute temperature scale starts here. The temperature of

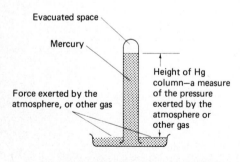

FIGURE 13-1 Simple barometer

absolute zero is −273.16° on the Celsius scale. In Fig. 13-2 we see a graphical comparison between the Celsius and the absolute, or Kelvin, temperature scales. We note that the size of the degree is the same on the two scales but the position of zero is different.

In making calculations using the gas law, the temperature that must be used is the *Kelvin temperature*. We can easily calculate this temperature from the Celsius temperature. To make this calculation, we add 273 to the Celsius reading. In summary,

$$°K = °C + 273$$

The value of R, the universal gas constant, has been determined experimentally, and it is practically the same regardless of the gas that is used in the experiment. Using the general gas law,

$$R = \frac{PV}{nT}$$

The units in which R is expressed depend on the units used for P and V. The value of n is always in moles, and T is always expressed in units of degrees Kelvin (°K). If pressure is expressed in atmospheres, and volume is expressed in liters, the units of R will be

$$\frac{\text{atmosphere} \times \text{liter}}{\text{mole} \times °K}$$

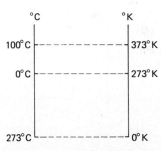

FIGURE 13-2 Comparison of the Celsius and Kelvin temperature scales

The numerical value of R may be calculated from experimental data as shown in the following example:

EXAMPLE Calculate the value of R from the following experimental data: One mole of hydrogen gas is found to occupy a volume of 22.4 liters at 0°C (or 273°K) and 1 atm (or 760 mm of Hg).

Step 1 Solving the gas law for R,

$$R = \frac{PV}{nT}$$

Step 2 Substitution of the above numbers into this expression gives

$$R = \frac{1 \text{ atm} \times 22.4 \text{ liters}}{1 \text{ mole} \times 273°K} = 0.0821 \frac{\text{atm} \times \text{liter}}{\text{mole} \times °K}$$

The conditions chosen in the previous example are defined as *standard conditions of temperature and pressure* (°C and 1 atm). Sometimes the abbreviation STP (standard temperature and pressure) is used to represent this set of conditions. These standard conditions define a convenient reference state.

You will note that there are five factors in the general gas law. If any four of these factors are known, the fifth one may be calculated. The value of R has been determined many times with many different substances and is found to have the constant value of 0.0821 liter × atmosphere/mole × °K. The following two examples illustrate the use of the general gas law:

EXAMPLE

A child's balloon is filled with helium gas. The volume of the balloon is 30.0 liters, and the pressure exerted on the gas is slightly greater than atmospheric pressure because of the elasticity of the rubber; it is measured to be 800 mm of Hg. The temperature of the gas is 27°C. Calculate the number of moles of helium contained in the balloon.

Step 1 Write the gas law: $PV = nRT$.

Step 2 List the information given in the problem:

$P = 800$ mm of Hg

$T = 27°C$

$V = 30.0$ liters

We also know that $R = 0.0821$ liter × atm/mole × °K. Since the units in which R is expressed here are liters, atmospheres, moles, and degrees Kelvin, these same units must be used to express V, P, n, and T.

Step 3 Change into units consistent with *R*:

$$P = (800 \text{ mm of Hg})\left(\frac{1 \text{ atm}}{760 \text{ mm of Hg}}\right) = 1.05 \text{ atm}$$

(*V* is already in liters)

$$T = 27 + 273 = 300°K$$

Step 4 Solve the general gas law for *n*: $PV = nRT$. Divide each side of the equation by RT: $PV/RT = n$.

Step 5 Substitute the numbers, and complete the arithmetic:

$$n = \frac{(1.05 \text{ atm})(30.0 \text{ liters})}{(8.21 \times 10^{-2} \text{ atm} \times \text{liter/mole} \times °K)(3.00 \times 10^2 °K)} =$$

1.28 moles of He

EXAMPLE Calculate the pressure exerted on the walls of a gas cylinder having a volume of 1.0 liter and containing 2.0 moles of nitrogen at 25°C.

Step 1 Write the general gas law: $PV = nRT$.

Step 2 List the information given in the problem:

$V = 1.0$ liter

$n = 2.0$ moles

$T = 25°C$

$R = 0.082 \dfrac{\text{liter-atm}}{°K\ \text{mole}}$

Step 3 Change the given information into units consistent with *R*:

$T = 25°C = 25 + 273 = 298°K$

(*n* is already in moles, and *V* is already in liters)

Step 4 Solve the general gas law for *P* by dividing both sides of the equation by *V*:

$P = \dfrac{nRT}{V}$

Step 5 Substitute the numbers and complete the arithmetic:

$P = \dfrac{\left(0.082 \dfrac{\cancel{\text{liter-atm}}}{\cancel{°K\ \text{mole}}}\right)(2.0\ \cancel{\text{moles}})(298°\cancel{K})}{2.0\ \cancel{\text{liters}}} = 2.45$ atm

13-3 CHANGES IN TEMPERATURE, PRESSURE, AND VOLUME

Common experience provides a qualitative understanding of the way the volume of a gas changes when temperature and pressure are changed. A familiar but dangerous example is placing an "empty" hair-spray can in an incinerator. As the temperature increases, the pressure increases, usually rupturing the can with an explosion when the container is no longer able to withstand the pressure. Also, if a balloon is cooled, the volume contracts since the pressure remains constant. These examples illustrate the effects of changing the factors in the general gas law.

If the number of moles of gas is kept constant (i.e., a certain quantity of gas is used) and the volume of the container remains constant, the pressure increases as the temperature increases, which is why the can explodes. In terms of the general gas law: $P = (nRT/V)$, where n, R, and V are all constant. P is observed to be directly proportional to T; in other words, the pressure increases as the temperature increases. If the absolute temperature is doubled, the pressure will also double. This is a statement of the *law of Gay-Lussac*.

A convenient mathematical expression which summarizes the concept that P is directly proportional to T may be stated as

$$\frac{P_1}{T_1} = \frac{P_2}{T_2}$$

where P_1 = original pressure
P_2 = final pressure

T_1 = original temperature, °K
T_2 = final temperature, °K

This equation may also be obtained from the general gas law by

$$P_1 V_1 = n_1 R T_1$$

for one set of conditions and

$$P_2 V_2 = n_2 R T_2$$

for another set of conditions. The first expression may be divided by the second expression to give

$$\frac{P_1 V_1 = n_1 R T_1}{P_2 V_2 = n_2 R T_2}$$

If the volume does not change, $V_1 = V_2$; if the number of moles of gas does not change, $n_1 = n_2$. And since R is always constant, V_1, n_1, and R all cancel in the above expression. This gives

$$\frac{P_1}{P_2} = \frac{T_1}{T_2} \quad \text{or} \quad \frac{P_1}{T_1} = \frac{P_2}{T_2}$$

EXAMPLE A hair-spray can has a fixed volume and contains a fixed number of moles of gas. Calculate the pressure that the gas inside the can will reach if the temperature changes from 27°C to 1500°K when the can is thrown into an incinerator. The initial pressure inside the can is 1.0 atm at 27°C.

Step 1 Write down the mathematical statement of the law of Gay-Lussac:

$$\frac{P_1}{T_1} = \frac{P_2}{T_2}$$

Step 2 List the information given in the problem:

$V_1 = V_2$ (the volume of the can does not change)

$n_1 = n_2$ (assuming the can is sealed, i.e., no gas enters or leaves the can)

$R = $ constant

$P_1 = 1$ atm

$T_1 = 27°C = 27 + 273 = 300°K$

$T_2 = 1500°K$

$P_2 = $ unknown

Step 3 Solve for P_2 by multiplying each side of the equation by T_2:

$$P_2 = \frac{P_1 \times T_2}{T_1}$$

Step 4 Substitute the numbers and complete the arithmetic:

$$P_2 = \frac{1 \text{ atm} \times 1500°K}{300°K} = 5 \text{ atm}$$

If we examine the logic in the preceding example, it is reasonable to expect pressure to increase as temperature increases; therefore P_1 should be multiplied by a factor larger than 1 to get P_2. As we learned previously, Boyle's law states that the volume of a gas is inversely proportional to the pressure if the temperature and number of moles remain constant. A mathematical statement of Boyle's law is

$$P_1 V_1 = P_2 V_2$$

This statement may also be obtained from the general gas law.

Charles' law, as you recall, states that the volume of a gas is directly proportional to the absolute temperature if the pressure and the number of moles of gas remain constant. A mathematical statement of Charles' law is

$$\frac{V_1}{T_1} = \frac{V_2}{T_2}$$

This equation may also be obtained from the general gas law.

EXAMPLE A weather balloon has an initial volume of 150.0 liters at 1.0 atm pressure. As it rises the pressure decreases to 0.80 atm. Calculate the final volume assuming constant temperature and no leaks in the balloon.

Step 1 Write the mathematical statement of Boyle's law:

$$P_1V_1 = P_2V_2$$

Step 2 List the information given in the problem:

$P_1 = 1.0$ atm

$P_2 = 0.80$ atm

$V_1 = 150.0$ liters

$V_2 =$ unknown

Step 3 Solve for V_2 by dividing each side by P_2:

$$\frac{P_1V_1}{P_2} = V_2$$

Step 4 Substitute the numbers and complete the arithmetic:

$$V_2 = \frac{1.0 \text{ atm} \times 150.0 \text{ liters}}{0.80 \text{ atm}} = 188.0 \text{ liters}$$

A balloon initially at 25°C has a volume of 25.0 liters. Calculate the volume the balloon will occupy if it is placed near a heater causing the temperature to rise to 50°C. Assume the pressure exerted on the gas is constant and there are no leaks in the balloon.

Step 1 Write the mathematical statement of Charles' law:

$$\frac{V_1}{T_1} = \frac{V_2}{T_2}$$

Step 2 List the information given in the problem:

$T_1 = 25°C = 25 + 273 = 298°K$

$V_1 = 25.0$ liters

$T_2 = 50°C = 50 + 273 = 323°K$

$V_2 =$ unknown

Step 3 Solve for V_2 by multiplying each side by T_2:

$$V_2 = \frac{T_2 \times V_1}{T_1}$$

Step 4 Substitute the numbers and complete the arithmetic:

$$V_2 = \frac{323°K \times 25.0 \text{ liters}}{298°K} = 27.1 \text{ liters}$$

In the two preceding examples, the algebra may be checked by considering logically whether the final volume will be increased or decreased by the change in pressure that occurs. That is, should V_1 be multiplied by the fraction 1.0/0.80, or should it be multiplied by the fraction 0.80/1.0? Since we know that volume increases as pressure decreases, the correct fraction is 1.0/0.80. Also, in the example dealing with Charles' law, we recognize that it is reasonable to expect the volume to increase as the temperature increases. Therefore, the initial volume is multiplied by the fraction 323°K/298°K.

In summary, the relationship

$$\frac{P_1 V_1}{T_1} = \frac{P_2 V_2}{T_2}$$

permits us to calculate any one of these factors if the other five are known. This *combined gas law* combines Charles' law ($V \, \alpha \, T$), Boyle's law ($P \, \alpha \, 1/V$), and Gay-Lussac's law ($P \, \alpha \, T$) into a single mathematical statement. (This type of calculation involves the implicit assumption that the number of moles of gas remains constant.)

EXAMPLE Carbon dioxide gas liberated from a blood sample occupies a volume of 3.00 ml at 25°C and a pressure of 750 mm of mercury. Compute the volume that the gas sample would occupy at STP.

Step 1 Write the mathematical statement of the combined gas law:

$$\frac{P_1 V_1}{T_1} = \frac{P_2 V_2}{T_2}$$

Step 2 List the information given in the problem:

$P_1 = 750$ mm

$T_1 = 25 + 273 = 298°K$

$V_1 = 3.00$ ml

$P_2 = 760$ mm

$T_2 = 273°K$

$V_2 =$ unknown

Step 3 Solve for V_2 by multiplying each side of the equation by T_2 and then dividing each side by P_2 to yield

$$V_2 = \frac{P_1 T_2 V_1}{P_2 T_1}$$

Step 4 Substitute the numbers and complete the arithmetic:

$$V_2 = \frac{750 \text{ mm} \times 273°K \times 3.00 \text{ ml}}{760 \text{ mm} \times 298°K} = 2.71 \text{ ml}$$

13-4 THE VOLUME OF 1 MOLE OF A GAS (MOLAR VOLUME)

It is often helpful to know the volume that is occupied by 1 mole of a gas because it enables the chemist to calculate the volume occupied by the gas generated in a chemical reaction. This volume has been determined by experiment to be 22.4 liters at STP. Let us consider the implications of this observation.

If we know the chemical formula of the gas that we are dealing with, we can calculate the weight of 1 mole of that gas. Since 1 mole of any gas at standard conditions occupies a volume of 22.4 liters, we can calculate the volume that will be occupied by any given weight of any particular gas at standard conditions. For example, suppose we are dealing with the gas carbon dioxide, which has the formula CO_2 and a molecular weight of 44.0 awu. Whenever 44.0 g of CO_2 gas are available, it will occupy 22.4 liters at STP.

13-5 WEIGHT-VOLUME PROBLEMS

Let us work through an example that makes use of a chemical equation to calculate the volume of gas that might be liberated as a result of a chemical reaction.

EXAMPLE Calculate the volume of carbon dioxide that will be liberated when 2.0 g of methane is burned. The equation for this reaction will be

$$CH_4 + 2CO_2 \rightarrow CO_2 + 2H_2O$$

Step 1 Calculate the moles of CH_4 initially available:

$$(2.0 \text{ g}) \left(\frac{1 \text{ mole}}{16 \text{ g}}\right) = 0.125 \text{ mole of } CH_4$$

Step 2 We know that one molecule of CO_2 is generated from one molecule of CH_4; therefore 0.125 mole of CO_2 will be formed.

Step 3 Calculate the volume of CO_2 at STP:

$$V \text{ of } CO_2 = \left(\frac{22.4 \text{ liters}}{1 \text{ mole}}\right) (0.125 \text{ mole}) = 2.8 \text{ liters}$$

In a similar manner, we can calculate the volume of gas liberated in any chemical reaction if we know the amount of one of the reactants and have a balanced equation describing the reaction that occurs.

GLOSSARY

Absolute zero: The lowest possible temperature ($-273.16°C$)

Barometer: A device for measuring the pressure of a gas

Boyle's law: Stated mathematically as $P_1 V_1 = P_2 V_2$

Charles' law: Stated mathematically as $V_1/T_1 = V_2/T_2$

Combined gas law: Stated mathematically as $P_1V_1/T_1 = P_2V_2/T_2$

Gay-Lussac's law: Stated mathematically as $P_1/T_1 = P_2/T_2$

General gas law: Stated mathematically as $PV = nRT$

Kelvin temperature scale (or the absolute temperature scale): A temperature scale starting at absolute zero using the same size degree as the Celsius scale (°K = 273 + °C)

Molar volume: The volume of one mole of gas at STP

Standard atmosphere: The pressure required to support a column of mercury 760 mm of height

Standard temperature and pressure (STP): 0°C (or 273°K) and 1 atm (or 760 mm of mercury)

Universal gas constant: $R = 0.0821$ atm × liters/°K × mole

SELF TEST

1. A device for measuring the pressure of a gas is called a _____.

2. The pressure of a gas may be expressed in _____ or _____.

3. The coldest possible temperature is called _____.

4. A temperature of 30°C is equal to _____ °K.

5. Give the mathematical expression for the general gas law.

6. Calculate the number of moles of gas in a cylinder that has a volume of 1.50 liters at a pressure of 2.3 atm and a temperature of 27°C.

7. The symbol R usually represents _____.

8. STP refers to a pressure of _____ and a temperature of _____.

9. For a sample of gas at constant temperature, the volume of the gas will _____ with an increase in pressure.

10. For a sample of gas at constant volume the pressure will _____ with an increase in the temperature.

11. The volume of 44.0 g of CO_2 is equal to _____ liters at STP and is called the _____.

12 If 1 mole of $CaCO_3$ (limestone) is decomposed by heating according to the equation

$$CaCO_3 \rightarrow CaO + CO_2$$

the volume of CO_2 liberated will be _____ at STP.

ANSWERS
1. Barometer
2. Atmospheres, millimeters of mercury
3. Absolute zero
4. 303°K
5. $PV = nRT$
6. 0.140 mole
7. The universal gas constant, or 0.082 liter-atm/°K mole
8. 760 mm (or 1 atm); 0°C (or 273°K)
9. Decrease
10. Increase
11. 22.4 liters, molar volume
12. 22.4 liters

EXERCISES
1. What is a barometer used for?

2   If the atmosphere supports a column of mercury 29.3 in. in length, what is the barometric pressure expressed in (a) millimeters of mercury and (b) standard atmospheres (1 in. = 25.4 mm)?

3   It is assumed that the pressure above the column of mercury inside a sealed glass tube that is inverted to make a barometer is ____0____ mm.

4   The three temperature scales that we have studied are ___C°___, ___F°___, and ___K°___.

5   Convert each of the following temperature readings to degrees absolute:
    (a) 10°C   (e) 150°C
    (b) −15°C   (f) 80°F
    (c) 100°C   (g) 40°F
    (d) 80°C   (h) 10°F

6   Solve the general gas law for
    (a) P   (d) R
    (b) V   (e) T
    (c) n

7   Calculate the pressure in a scuba tank with a volume of 14.0 liters if the temperature is 35°C and the tank contains a total of 4.0 moles of gas.

8   A diver submerges from the surface where the temperature is 20°C to a depth where the water temperature is 10°C. Assuming the initial scuba-tank pressure is 100 atm, compute the tank pressure at the new temperature.

9   Compute the volume occupied by 10.0 g of helium gas at 25°C and 1.0 atm pressure. 61.2 l

10 Calculate the volume occupied by 32.0 g of methane (or natural gas, CH_4) at 1.5 atm and 30°C.

11 Compute the number of moles of natural gas present in a container having a volume of 30,000 liters at a temperature of 22°C and a pressure of 1.2 atm.

12 Calculate the number of moles of air in a basketball whose volume is 25 liters at 30°C and 1.8 atm pressure. (*Hint:* The number of moles of gas does not depend on the chemical formula of the gas.)
     1.81 mol

13 Calculate the temperature generated in the cylinder of a diesel engine if the pressure becomes 50.0 atm, the number of moles of gas present is 0.03, and the volume is 100.0 ml.

14 What is the temperature of gas in a cylinder if the gas pressure is 200.0 atm, the volume of the cylinder is 50.0 liters, and it contains 300 moles of gas?

15 For a sample of gas at constant temperature the volume of the gas will ___↗___ with a decrease in pressure.

16 For a sample of a gas at constant temperature the volume of the gas will increase with a(n) _____ in pressure.

17 For a sample of a gas at constant volume the pressure will increase with a(n) _____ in temperature.

18 If the temperature of a gas increases, its volume will ___↗___ at constant pressure.

19 If the temperature of a gas _____, its volume will decrease at constant pressure.

20 Write the mathematical statement of
(a) Boyle's law   (b) Charles' law   (c) Gay-Lussac's law

21 Calculate the volume that will be occupied by 100 cc of a gas if its pressure changes from 760 to 800 mm with the temperature remaining constant.   94.8 cc

22 If a balloon having a volume of 2 liters at a pressure of 900 mm and temperature of 0°C is placed under standard conditions, what volume will it occupy?

23. If the volume of a certain sample of gas at 0.50 atm is changed from 6 to 8 liters while the temperature remains constant, calculate the new gas pressure.

24. If 25 ml of oxygen has its pressure increased from 15 to 20 lb/in.^{2} while the temperature remains constant, what volume will it occupy?

25. In each of the following examples, assume the pressure remains constant and the indicated changes in temperature occur. In each case calculate the final volume of the gas:
    (a) 10 liters of carbon dioxide has its temperature changed from 20 to 50°C.
    (b) 250 cc of oxygen are heated from 0 to 300°C.
    (c) 1 pt of hydrogen is cooled from 300 to −10°C.
    (d) 2 liters of chlorine are changed from 100°C to standard temperature.
    (e) 2 qt of oxygen are heated from 15 to 100°F.

26. In each of the following examples, calculate the volume that would be occupied by the gas if it were changed to STP:
    (a) 25 liters of oxygen at 20°C and 650 mm
    (b) 540 ml of methane at 45°C and 900 mm
    (c) 45 ml of methane at −10°C and 800 mm
    (d) 300 cc of chlorine at 0°C and 1,000 mm
    (e) 175 ml of nitrogen at 25°F and 850 mm

27. Calculate the number of moles of gas in each example in Exercise 26.

28. Calculate the weight of gas liberated and the volume that it will occupy at STP in each example:
   (a) 12 g of methane (CH_4) are burned. (Two gases are formed here—water and carbon dioxide; make the calculation for each one.)
   (b) 3 g of potassium chlorate decompose. The equation for the reaction is

   $$2KClO_3 \xrightarrow{\Delta} 2KCl + 3O_2$$

   (c) 15 g of mercuric oxide decompose. The equation for the reaction is

   $$2HgO \rightarrow 2Hg + O_2$$

29. If we have 5 g of $KClO_3$ in a 1.0-liter evacuated container and heat this material until it all decomposes, calculate the pressure that will be exerted by the gas if the temperature is 1000°C. The equation for the reaction is

   $$2KClO_3 \rightarrow 3O_2 + 2KCl$$

30. Compute the volume of CO_2 formed at STP if 30 g of butane (C_4H_{10}) are burned. The equation for the reaction is   46.3 l

   $$2C_4H_{10} + 13O_2 \rightarrow 8CO_2 + 10H_2O$$

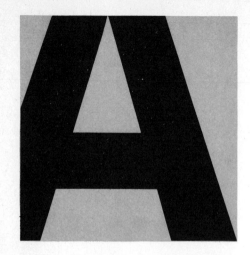

APPENDIXES

I-1 EXPRESSING NUMBERS IN EXPONENTIAL NOTATION

APPENDIX I
Exponential Notation
(Scientific Notation)

It is frequently desirable to express a number using powers of 10. This is most often useful when dealing with very large or very small numbers. Table I-1 shows various numbers written in the usual manner and also written using exponential notation (powers of 10).

TABLE I-1 Exponential notation

NUMBER	EXPONENTIAL NOTATION
Numbers greater than 1	
1,000,000	1×10^6
100,000	1×10^5
10,000	1×10^4
1,000	1×10^3
100	1×10^2
10	1×10^1
1	1×10^0
Numbers between 0 and 1	
0.10	1×10^{-1}
0.010	1×10^{-2}
0.0010	1×10^{-3}
0.00010	1×10^{-4}
0.000010	1×10^{-5}
0.0000010	1×10^{-6}

EXAMPLES

$454 = 4.54 \times 100 = 4.54 \times 10 \times 10 = 4.54 \times 10^2$

$5{,}280 = 5.280 \times 1{,}000 = 5.280 \times 10 \times 10 \times 10 = 5.280 \times 10^3$

$1{,}786{,}537 = 1.786537 \times 10^6$

$602{,}000{,}000{,}000{,}000{,}000{,}000{,}000 = 602 \times 10^{21} = 6.02 \times 10^{23}$

$0.105 = \dfrac{1.05}{10} = 1.05 \times 10^{-1}$

$0.000782 = \dfrac{7.82}{10{,}000} = 7.82 \times 10^{-4}$

$0.00000000000000000000000166 = 1.66 \times 10^{-24}$

You will note that in order to express a number greater than 1 using exponential notation, you must add 1 to the exponent for each place the decimal is moved to the left. For numbers less than 1, you must subtract 1 from the exponent for each place the decimal is moved to the right.

APPENDIX II
Algebraic Manipulations APPENDIXES

II-1 ADDITION AND SUBTRACTION USING POWERS OF 10
When adding and subtracting numbers written as powers of 10, it is always necessary to be sure the decimal point is properly placed.

EXAMPLES

$1.07 \times 10^3 + 2.78 \times 10^2 = 10.7 \times 10^2 + 2.78 \times 10^2$

$$\begin{array}{r} 10.7 \times 10^2 \\ +\ \ 2.78 \times 10^2 \\ \hline 13.48 \times 10^2 \end{array}$$

In order to add numbers correctly, the initial numbers must be multiplied by the same power of 10 and the decimal points must be in the same column. In subtraction, the same rules must be used. Note that the power of 10 serves only to place the decimal in the number.

II-2 MULTIPLICATION
In multiplying numbers written as powers of 10, the numbers preceding the powers of 10 are multiplied and the exponents are added.

EXAMPLES

$(1.78 \times 10^3) \times (2.85 \times 10^2) = 5.07 \times 10^5$

$(2.37 \times 10^{-3}) \times (8.76 \times 10^2) = 20.7 \times 10^{-1} = 2.07$

II-3 DIVISION

In dividing numbers written as powers of 10, the numbers preceding the powers of 10 are divided and the exponents are subtracted.

EXAMPLES

$$(1.87 \times 10^3) \div (2.76 \times 10^2) = \frac{1.87 \times 100^3}{2.76 \times 10^2} \quad \text{or } 0.678 \times 10^1 = 6.78$$

$$(9.76 \times 10^6) \div (2.35 \times 10^{-2}) = 4.16 \times 10^8$$

II-4 COMBINED OPERATIONS

In a series of operations involving several different manipulations, it is necessary to work out the problem in a stepwise fashion.

EXAMPLE Solve for X:

$$\frac{1.5 + 2.8}{2.3 \times 10^2} = \frac{X}{1.3 \times 10^{-1}}$$

$$\frac{4.3}{2.3 \times 10^2} = \frac{X}{1.3 \times 10^{-1}} \qquad \text{necessary addition and subtraction is carried out first}$$

$$\frac{(1.3 \times 10^{-1}) \times 4.3}{2.3 \times 10^2} = X \qquad \text{both sides are multiplied by } 1.3 \times 10^{-1}$$

$$\frac{5.59 \times 10^{-1}}{2.3 \times 10^2} = X \qquad \text{multiplication is performed}$$

$$2.4 \times 100^{-3} = X \qquad \text{division is performed}$$

APPENDIX III
Conversion Factors

	Conversion to Metric Units	Conversion to English Units
Length Equivalents		
1 inch = 2.54 centimeters	2.54 cm/1 in.	1 in./2.54 cm
0.394 in. = 1.00 cm	1.00 cm/0.394 in.	0.394 in./1.00 cm
39.4 in. = 1.00 meter	1.00 m/39.4 in.	39.4 in./1.00 m
1 mile = 1.61 kilometers	1.61 km/1 mile	1.0 mile/1.61 km
Volume Equivalents		
1 pint = 0.473 liters	0.473 liter/1 pt	1 pt/0.473 liter
1.06 quarts = 1.00 liters	1.00 liter/1.06 qt	1.06 qt/1.00 liter
1 gallon = 3.79 liters	3.79 liters/1 gal	1 gal/3.79 liters
Weight (or mass): Avoirdupois Equivalents		
0.0353 ounces = 1 gram	1 g/0.0353 oz	0.0353 oz/1 g
1 oz = 28.3 g	28.3 g/1 oz	1 oz/28.3 g
1 pound = 454 g	454 g/1 lb	1 lb/454 g
2.20 lb = 1 kilogram	1 kg/2.20 lb	2.20 lb/1 kg
Miscellaneous Equivalents		
1 angstrom = 3.94×10^{-9} in. (or 1×10^{-8} cm)	1 Å/3.94×10^{-9} in.	3.94×10^{-9} in./1 Å
1 carat = 200 milligrams (or 3.087 grains)	200 mg/1 carat	1 carat/200 mg
1 stone = 6.35 kg (14.0 lb)	6.35 kg/1 stone	1 stone/6.35 kg

Miscellaneous Equivalents (Continued)

1 cubit = 45.72 cm (or 18 in.)	45.72 cm/1 cubit	1 cubit/45.72 cm
1 fathom = 1.83 m (or 6 feet)	1.83 m/1 fathom	1 fathom/1.83 m
1 gr = 0.0648 g (0.00229 oz)	0.0648 g/1 gr	1 gr/0.0648 g
1 ton = 907 kg (or 2000 lb)	907 kg/1 ton	1 ton/907 kg

APPENDIX IV
Electronic Configuration of the First 36 Elements

The electrons in atoms are arranged in shells and subshells. The major shells are designed K, L, M, N, etcetera. The subshells are designated s, p, d, and f. The major shells lie at increasing distances (energies) from the nucleus of the atom. The subshells, composed of orbitals, represent the regions in space where the electrons within a particular subshell are most likely to be located. The relative energies of these shells and subshells are diagramed in Table IV-1. The 2s and 2p sublevels make up the L shell, or the second main energy level; also, the 3s, 3p, and 3d sublevels make up the M shell, or third main energy level.

Each orbital may contain a maximum of two electrons. Thus we note in Table IV-2 that the 1s sublevel may contain a maximum of two electrons and the 2p sublevel may contain a maximum of six electrons (in three orbitals). This plus the two 2s electrons makes up the eight electrons possible in the L shell. As an example of the interpretation of Table IV-2, the element iron,

TABLE IV-1 Energy sublevels

atomic number 26, has an electronic configuration that can be represented as $1s^2 2s^2 2p^6 3s^2 3p^6 3d^6 4s^2$. The symbol $2p^6$ here, for example, means that in the second main shell of electrons there are six electrons within the $3p$ orbitals.

The relationship between Table IV-2 and the periodic chart (Table 2-4 in the text, page 44 shows that all those elements having one electron in the s orbital of the valence shell fall in group IA; all those elements with two electrons in the s orbital of the valence shell *and* two electrons in the p orbitals of the valence shell fall in group IVA; and so forth.

TABLE IV-2 Electronic configuration of the first 36 elements

ATOMIC NO.	ELEMENT	K 1 s	L 2 s	L 2 p	M 3 s	M 3 p	M 3 d	N 4 s	N 4 p	N 4 d	N 4 f	MAIN ENERGY LEVELS ORBITALS
1	H	1										
2	He	2										
3	Li	2	1									
4	Be	2	2									
5	B	2	2	1								
6	C	2	2	2								
7	N	2	2	3								
8	O	2	2	4								
9	F	2	2	5								
10	Ne	2	2	6								
11	Na	2	2	6	1							
12	Mg	2	2	6	2							

TABLE IV-2 Electronic configuration of the first 36 elements (*Continued*)

ATOMIC NO.	ELEMENT	K 1	L 2		M 3			N 4			MAIN ENERGY LEVELS
		s	s	p	s	p	d	s	p	d	f	ORBITALS
13	Al	2	2	6	2	1						
14	Si	2	2	6	2	2						
15	P	2	2	6	2	3						
16	S	2	2	6	2	4						
17	Cl	2	2	6	2	5						
18	Ar	2	2	6	2	6						
19	K	2	2	6	2	6		1				
20	Ca	2	2	6	2	6		2				
21	Sc	2	2	6	2	6	1	2				
22	Ti	2	2	6	2	6	2	2				
23	V	2	2	6	2	6	3	2				
24	Cr	2	2	6	2	6	5	1				
25	Mn	2	2	6	2	6	5	2				
26	Fe	2	2	6	2	6	6	2				
27	Co	2	2	6	2	6	7	2				
28	Ni	2	2	6	2	6	8	2				
29	Cu	2	2	6	2	6	10	1				
30	Zn	2	2	6	2	6	10	2				
31	Ga	2	2	6	2	6	10	2	1			
32	Ge	2	2	6	2	6	10	2	2			
33	As	2	2	6	2	6	10	2	3			
34	Se	2	2	6	2	6	10	2	4			
35	Br	2	2	6	2	6	10	2	5			
36	Kr	2	2	6	2	6	10	2	6			

APPENDIX V
Shapes of Orbitals and Molecules

V-1 SHAPES OF ORBITALS

An *orbital* is a region in space within an atom where not more than two electrons may reside. The shapes of orbitals have been calculated by methods of quantum mechanics, and they are represented in Fig. V-1.

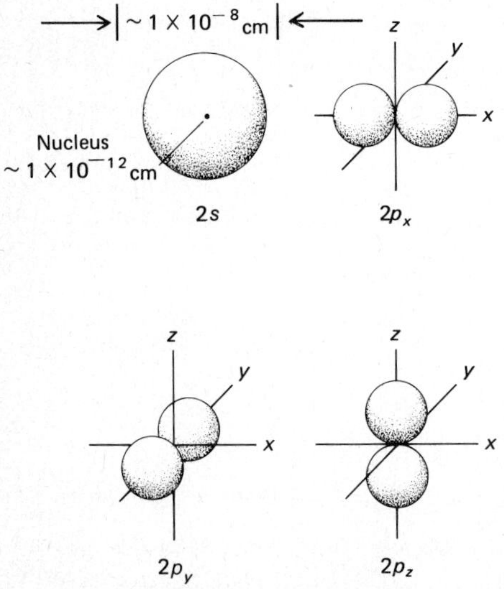

V-2 SHAPES OF MOLECULES

Logical reasoning leads us to reasonable predictions of the shapes of simple molecules. For example, a molecule composed of two atoms (diatomic) *must be linear* (see Fig. V-2). A molecule composed of three atoms (triatomic) may

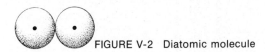

FIGURE V-2 Diatomic molecule

either be linear or angular. A prediction as to a linear or angular molecule may be made by considering the orbital distribution of electrons. Electrons tend to distribute themselves in pairs in orbitals, and these pairs tend to be as far apart as possible since negative particles repel each other.

For example, in water there are four pairs of electrons around the central oxygen atom (see Fig. V-3). These four pairs of electrons distribute them-

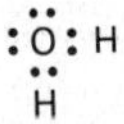

FIGURE V-3 Electron-dot structure of a water molecule

FIGURE V-1 Shapes of orbitals

selves in four orbitals which are as far apart as possible; thus they point to the four corners of a regular tetrahedron (see Fig. V-4). Placing the two hydrogen atoms at the corners of the tetrahedron produces an angular molecule (see Fig. V-5). The covalent bonds between the oxygen and hydrogen atoms are formed by the overlap in space of the orbitals of the oxygen atom with the orbitals of the hydrogen atom.

In carbon dioxide the electrons are distributed in the manner shown in Fig. V-6.

$\ddot{\text{O}}::\text{C}::\ddot{\text{O}}$ FIGURE V-6   Electron-dot structure of CO_2

The two double bonds to the carbon atom are placed as far apart as possible, thus producing a linear molecule. The shapes of more complex molecules may sometimes be predicted by combining the shapes of the smaller portions of the molecules

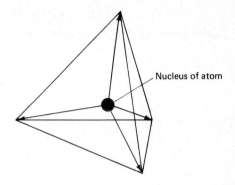

FIGURE V-4   Tetrahedral arrangement

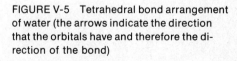

FIGURE V-5   Tetrahedral bond arrangement of water (the arrows indicate the direction that the orbitals have and therefore the direction of the bond)

APPENDIX VI
Additional Inorganic Nomenclature

VI-1 NOMENCLATURE OF OXIDES

When a nonmetallic element forms many different compounds with oxygen, the number of atoms of each element may be indicated by the prefixes mono, di, tri, and so on. As examples, let us consider the nomenclature of the oxides of nitrogen:

N_2O	Dinitrogen monoxide (or nitrous oxide)
NO	Nitrogen monoxide (or nitric oxide)
N_2O_3	Dinitrogen trioxide
NO_2	Nitrogen dioxide
N_2O_4	Dinitrogen tetroxide
N_2O_5	Dinitrogen pentoxide

Similar nomenclature is used for phosphorus, sulfur, and other nonmetallic oxides.

TABLE VI-1 Names of additional acids, polyatomic ions, and other ions

ACIDS		POLYATOMIC IONS	
$HClO_4$	Perchloric acid	ClO_4^-	Perchlorate ion
$HClO_3$	Chloric acid	ClO_3^-	Chlorate ion
$HClO_2$	Chlorous acid	ClO_2^-	Chlorite ion
$HClO$	Hypochlorous acid	ClO^-	Hypochlorite ion
H_3BO_3	Boric acid	BO_3^{3-}	Borate ion
H_2SiO_3	Silicic acid	SiO_3^-	Silicate ion
H_3PO_3	Phosphorous acid	PO_3^{3-}	Phosphite ion

TABLE VI-1   Names of additional acids, polyatomic ions, and other ions (*Continued*)

HNO_2	Nitrous acid	NO_2^-	Nitrite ion
H_2SO_3	Sulfurous acid	SO_3^{2-}	Sulfite ion

OTHER IONS

$H_2PO_4^-$	Dihydrogen phosphate
HPO_4^{2-}	Hydrogen phosphate
HCO_3^-	Hydrogen carbonate (bicarbonate)
HSO_4^-	Hydrogen sulfate (bisulfate)
MnO_4^-	Permanganate
$Cr_2O_7^{2-}$	Dichromate
$C_2O_4^{2-}$	Oxalate
O_2^{2-}	Peroxide
AsO_4^{3-}	Arsenate
AsO_3^{3-}	Arsenite

APPENDIX VII
Answers to Selected Exercises

ANSWERS CHAPTER 1

3 See Glossary.
6 (b) 10^3 mm/1 m; (c) 10^2 cm/1 m; (d) 1 m/10^2 cm; (e) 10^5 cm/1 km; (f) 1 liter/1.06 qt; (g) 1.06 qt/1 liter; (h) 10^3 ml/1 liter; (i) 1 kg/10^3 g; (j) 10^6 mg/1 kg
9 1094 yd
12 (a) 2.12 pt; (b) 188.7 cc
15 $-321°F$
18 $-108.4°F$
21 1.81 ml
24 74.7 lb
27 1.00 μm
30 9.27×10^{-2} oz
33 705 lb/ft^3
36 (a) 6.44×10^{14} g; (b) 6.44×10^{11} kg; (c) 1.42×10^{12} lb

ANSWERS CHAPTER 2

3 See Glossary.
6 See Table 2-2.
9 About 50°C
12 See Glossary.
15 Not exactly, but some properties would be similar since Na and K are both in the same group in the periodic table (Table 2-4).

18   Magnesium, manganese, sodium, potassium, chlorine, gold, bromine, beryllium, helium, tungsten, iron, silver, lead, sulfur, and phosphorus

ANSWERS CHAPTER 3

3   See Glossary.
6   (a) 1 to 5 Å; (b) 1×10^{-7} to 5×10^{-7} mm; (c) 1×10^{-8} to 5×10^{-8} cm; (d) 3.94×10^{-9} to 1.97×10^{-8} in.
9   Bohr's theory is somewhat analogous to the solar system, the planets being analogous to electrons and the sun to the nucleus.
12   Yes, an electron is a fundamental particle.
15   (a) IIA; (b) two; (c) 18; (d) 20; (e) 20; (f) +2; (g) Ar
18   55.8 g
21   See Fig. 3-9.
24   For members of the A groups, the number of the period gives the number of the main energy level in which the valence electrons are located.
27   The atomic number gives the number of protons in the nucleus of the atom; it is also equal to the number of electrons present in the atom.
30   (a) $^{35}_{17}Cl$; (b) $^{37}_{17}Cl$; (c) $^{12}_{6}C$; (d) $^{14}_{6}C$ (e) $^{238}_{92}U$

ANSWERS CHAPTER 4

3   No, since sodium chloride is a compound, a chemical change would have to take place to separate the chlorine from the sodium.

6   1,117.0 lb
9   (a) VIIA; (b) seven; (c) −1
12 (a) VIIA; (b) seven; (c) −1
15 (a) $(:\!\ddot{\underset{..}{Cl}}\!:)^-$; (b) $(:\!\ddot{O}\!:\!H)^-$; (c) $(Li)^+$ because the lone valence electron has been lost; (d) $(Ca)^{++}$; (e) $:\!\ddot{\underset{..}{Ne}}\!:$; (f) $(:\!\ddot{\underset{..}{S}}\!:)^{--}$; (g) $:\!\ddot{\underset{..}{Cl}}\!:\!\ddot{\underset{..}{Cl}}\!:$

(h) $\left[\begin{array}{c} :\ddot{O}: \\ :\ddot{O}:P:\ddot{O}: \\ :\ddot{O}: \end{array}\right]^{3-}$ 　　(i) $\left[\begin{array}{c} :\ddot{O}:N:\ddot{O}: \\ :: \\ :\ddot{O}: \end{array}\right]^{-}$

18 (a) Ionic
    (b) Ionic
    (c) Ionic
    (d) Covalent
    (e) Covalent
    (f) Ionic
    (g) Covalent
    (h) Covalent

ANSWERS   CHAPTER 5

3   Iron, tin, and lead can show more than one electrovalence. Going left to right across the last three rows of the table in Exercise 1, the names are iron(III) nitrate, iron(III) acetate, iron(III) sulfate, iron(III) carbonate,

iron (III) phosphate; tin(II) nitrate, tin(II) acetate, tin(II) sulfate, tin(II) carbonate, tin(II) phosphate; lead(IV) nitrate, lead(IV) acetate, lead(IV) sulfate, lead(IV) carbonate, and lead(IV) phosphate.

6 (a) Mercury(II) chloride (k) Plumbous sulfate
   (b) Mercury (I) bromide   (l) Lead(II) carbonate
   (c) Lead(II) oxide   (m) Mercuric sulfate
   (d) Lead(IV) oxide   (n) Mercurous sulfate
   (e) Lead(II) chloride   (o) Copper(II) hydroxide
   (f) Mercury(II) sulfide   (p) Iron(III) sulfate
   (g) Copper(II) sulfide   (q) Stannous carbonate
   (h) Copper(I) sulfide   (r) Iron(II) carbonate
   (i) Ferrous bromide   (s) Stannic phosphate
   (j) Ferric oxide   (t) Tin(II) nitrate

9 (a) Na· + :Cl̈: → Na⁺ + :C̈l:⁻

   (b) Na· + :Ö: → Na⁺ + :Ö:⁻⁻
       Na·   Na⁺

   (c) Mg· + :C̈l: → Mg⁺⁺ + :C̈l:⁻
           :C̈l: :C̈l:⁻

   (d) Ca· + :Ö: → Ca⁺⁺ + :Ö:⁻⁻

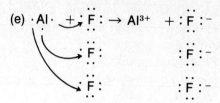

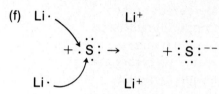

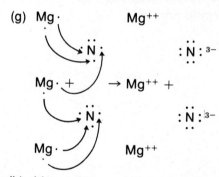

(h) Li· + ;H → Li⁺ + :H⁻

12 (a) Monoxide   (e) Trioxide
    (b) Dioxide   (f) Trioxide
    (c) Tetroxide   (g) Trioxide
    (d) Pentoxide

ANSWERS CHAPTER 6

3. See Glossary.
6. (a) 83.7% Cl and 16.3% Li; (b) 20.1% Ca and 79.9% Br; (c) 37.5% C, 12.6% H, and 50.0% O; (d) 60.3% Mg and 39.7% O; (e) 26.1% N, 7.55% H, and 66.3% Cl; (f) 52.9% Al and 47.1% O; (g) 46.8% Si and 53.2% O; (h) 39.9% K, 24.5% C, 3.09% H, and 32.6% O; (i) 35.0% N, 5.05% H, and 60.0% O; (j) 15.8% Al, 28.1% S, and 56.1% O.
9. (a) 65.2
   (b) 1.0 mole
   (c) 1.0 mole
   (d) 4.0 moles
   (e) $ZnSO_4$
12. (a) 48.0 awu
    (b) 0.521 mole
    (c) 1.56 moles
    (d) 153.6 g
    (e) 32.0 g
15. See Glossary.
18. (a) 18.1×10^{23} (g) 7.47×10^{22}
    (b) 6.02×10^{22} (h) 2.84×10^{23}
    (c) 4.82×10^{23} (i) 2.43×10^{23}
    (d) 1.20×10^{23} (j) 1.00×10^{23}
    (e) 1.20×10^{24} (k) 5.74×10^{21}
    (f) 1.20×10^{23}
21. (a) $C_3H_4O_3$
    (b) $C_6H_8O_6$

ANSWERS CHAPTER 7

3. See Table 7-1.
6. (a) Cl_2; (b) O_2; (c) HCl; (d) $ZnSO_4$; (e) $NaNO_3$; (f) O_2; (g) Na_2CO_3; (h) $Cu(NO_3)_2$; (i) $Zn(NO_3)_2$; (j) Ag
9. (a) Al_2O_3; (b) PbS, $K(NO_3)$; (c) CO_2, H_2O; (d) H_2O, $Ca_3(PO_4)_2$; (e) CO_2; (f) $Al(NO_3)_3$, H_2O; (g) CaO; (h) NH_3; (i) $Zn(C_2H_3O_2)_2$, H_2; (j) HgS, $(NH_4)(NO_3)$

ANSWERS CHAPTER 8

3. 1,000 molecules of CO_2 and 2,000 molecules of H_2O
6. $19\frac{1}{2}$; 12; 15; $\frac{1}{13}$; $\frac{4}{13}$; $\frac{5}{13}$; 35.9; 0.69; 0.86

9. $Ca + Cl_2 \rightarrow CaCl_2$
   (a) 40.1; 71; 111.1
   (b) 17.7 g
   (c) 27.7 g
   (d) 15.69

12. (a) 0.10 mole
    (b) 7.1 g
    (c) 7.1 g
    (d) 0.4 mole of HCl is 14.6 g

15. (a) 1,340 kg of SO_2
    (b) 2,050 kg of $H_2(SO_4)$, or 2.05×10^6 g of H_2SO_4

18. 15.3 g of alcohol

21 (a) $NaHCO_3$ is the limiting reagent.
    (b) $H(C_2H_3O_2)$ is present in excess.
    (c) 0.25 mole of CO_2 is formed.
    (d) 11 g
    (e) 0.25 mole of H_2O is formed.
    (f) 4.5 g
    (g) 15 g

ANSWERS CHAPTER 9
3   Solid
6   See Glossary.
9   See text.
12 See Glossary.
15 Rapidly; vibrate, rapidly; straight lines; colliding; leave
18 It will probably explode as a result of increasing pressure.

ANSWERS CHAPTER 10
3   1 ~~lb/gal~~ × 454 g/~~lb~~ × 1 ~~gal~~/3.79 ~~liter~~ × 1 ~~liter~~/1,000 ml = 0.12 g/ml
6   (a) 6 g/liter
    (b) 0.061 mole/liter
9   0.00125 liter, or 1.25 ml
12 0.53 mole/liter

15 185 g
18 0.061 M
21 6.08 M
24 (a) 1.26 M
   (b) 37.4 g Na(NO$_3$)
27 83.4 ml
30 (a) 3.9 × 10^{-3}%
   (b) 0.00122 M

ANSWERS CHAPTER 11

3 OH$^-$
6 See Fig. 11-1.
9 H$^+$
12 (a) HCl + NaOH → NaCl + HOH
   (b) H$_2$SO$_4$ + 2NaOH → Na$_2$SO$_4$ + 2HOH
   (c) H$_3$PO$_4$ + 3LiOH → Li$_3$PO$_4$ + 3HOH
   (d) 2HC$_2$H$_3$O$_2$ + Ca(OH)$_2$ → Ca(C$_2$H$_3$O$_2$)$_2$ + 2HOH
   (e) 2HNO$_3$ + Ca(OH)$_2$ → Ca(NO$_3$)$_2$ + 2HOH
   (f) 3HNO$_3$ + Al(OH)$_3$ → Al(NO$_3$)$_3$ + 3HOH
   (g) 2H$_3$PO$_4$ + 3Ca(OH)$_2$ → Ca$_3$(PO$_4$)$_2$ + 6HOH
   (h) H$_3$PO$_4$ + Al(OH)$_3$ → AlPO$_4$ + 3HOH
15 2.25 g
18 (a) MgCO$_3$ + 2HCl → MgCl$_2$ + H$_2$CO$_3$ followed by
       H$_2$CO$_3$ → H$_2$O + CO$_2$ ↑

(b) The reaction tends to convert any acid present into CO_2, H_2O, and a salt.

(c) The CO_2 gas produced needs to escape from the stomach.

21 $3\ M$

24 (a) $Ba^{++} + \cancel{2NO_3^-} + \cancel{2Na^+} + SO_4^{--} \rightarrow BaSO_4 + \cancel{2Na^+} + \cancel{2NO_3^-}$

(b) $Zn^{++} + \cancel{2Cl^-} + \cancel{2NH_4^+} + S^{--} \rightarrow ZnS\downarrow + \cancel{2NH_4^+} + \cancel{2Cl^-}$

(c) $2Ag^+ + \cancel{2NO_3^-} + \cancel{2Na^+} + CO_3^{--} \rightarrow Ag_2CO_3\downarrow + \cancel{2Na^+} + \cancel{2NO_3^-}$

(d) $\cancel{6Na^+} + 2PO_4^{3-} + 3Ca^{++} + \cancel{6NO_3^-} \rightarrow Ca_3(PO_4)_2\downarrow + \cancel{6Na^+} + \cancel{6NO_3^-}$

(e) $Pb^{++} + \cancel{2NO_3^-} + \cancel{2Na^+} + 2Cl^- \rightarrow PbCl_2\downarrow + \cancel{2Na^+} + \cancel{2NO_3^-}$

27 (a) $PbCl_2$

(b) CdS

(c) $CaCO_3$

30 0.003 mole of H^+

33 $0.500\ N$

36 $26\frac{2}{3}$ ml

39 (a) $0.75\ N$

(b) $0.25\ M$

ANSWERS CHAPTER 12

3 The equilibrium can be represented by $H_2CO_3 \rightleftharpoons H_2O + CO_2$. When the cap is removed, the CO_2 escapes from the bottle. According to the law of Le Châtelier, the removal of CO_2 would cause the above equilibrium to shift to the right. Eventually all the CO_2 would escape, leaving the softdrink "flat," or uncarbonated.

6 The sodium ions (and chloride ions) are going from the solid NaCl into solution and from solution back to the solid NaCl at the same rate. The weight of undissolved NaCl does not change nor does the concentration of sodium chloride in the solution.
9 A strong acid is one which is completely ionized in a water solution.
12 (a) strong; (b) strong; (c) weak; (d) weak
15 HCl
18 $3 \times 10^{-5}\ M$
21 (a) left; (b) decrease; (c) increase
24 $pH = -\log(H^+)$
27 3 since HNO_3 is a strong acid and completely ionized

ANSWERS CHAPTER 13

3   0 mm
6   (a) $P = nRT/V$; (b) $V = nRT/P$; (c) $n = PV/RT$; (d) $R = PV/nT$; (e) $T = PV/nR$
9   61.2 liters
12   1.81 moles
15   Increase
18   Increase
21   94.8 cc
24   18.8 ml
27   (a) 0.888 mole; (b) 2.45×10^{-2} mole; (c) 2.19×10^{-3} mole; (d) 1.76×10^{-2} mole; (e) 8.87×10^{-2} mole
30   46.3 liters

Page numbers in *italic* indicate glossary definitions.

INDEX

Absolute zero, *297*
Acid, 236–239, *249*
Acid salt, 243, *249*
Algebraic manipulations, 309, 310
Alkali metals, 46, *47*
Alloy, 215, *226*
Alpha particles, 60
Alumina, 108
Ammonium chloride, 93
   structure, 93
Amorphous solid, 200, *207*
Arrhenius theory of ionization, 119
Atom, 56, *72*
Atomic number, 45, 63, 69, *72*
Atomic weight, 45
Atomic weight unit, 62, 67, 68, *72, 138*
Avogadro's number, 123, *138*

Balancing a chemical equation, 151–155
Balancing coefficient, *184*
Barometer, *297*
Base, 239, *250*
Binary compound, 102, 103, *109*, 238, *250*

Boiling point, 205, 206, *207*
Boyle's law, 202, 203, *207*, 291, *297*
Brownian motion, 57, *72*

Carbon dioxide, structure, 87
Cathode rays, 59, *72*
Celsius temperature, 18, *26*
Centimeter, 10
Charles' law, 202, 204, *207*, 291; *298*
Chemical equation, 150–152, *161*
Chemical nomenclature, 102–109
Chemical reaction, 148–149, *161*
   types of, 152, 153
Chemistry, 3, *26*
Chlorine, properties, 83
Cloud chamber, 57
Combination reaction, *161*
Combined gas law, 294, *298*
Combustion reaction, 152, 159, *161*
Compound, 5, *27*, 82, 83, *94*
Concentration, 215–217, *227*
Conservation of matter, 55, *73*
Conversion factor, 14, 311, 312
Conversion of units, 14

Covalence, 86, 87, 91, *94*
Covalent bond, 86, 87, *94*
Covalent molecules, 119
Crystal lattice, 199, 200, *207*

Dalton, John, 55
Dalton's atomic theory, 55, *73*
Decomposition reaction, *161*
Density, *27*
  units of, 22
Diatomic elements, 150
Dilution, 221, *227*
Dimensional analysis, 16, *27*
Displacement reaction, *161*
Double bond, 87, *94*
Dynamic equilibrium, 262–263, *270*

Ekasilicon, 45
Electrolysis, *162*
Electron, 59, 60, 63, 66, 67, *73*
Electron configuration of elements, 314, 315
Electron-dot structure, 86–89, *94*
Electron shell, 65–67, 70–72, *73*
Electrovalence, 90, 91, *94*

Element, 4, *27*, 37
Elements:
  common, 40–43
  man-made, 38, 41
Empirical formula, 130–138, *138*
Energy sublevels, 313
English system of units, 8
Equilibrium, 262–278
Equivalent:
  of an acid, 247, *250*
  of a base, 247–248, *250*
Evaporation, 205, *207*
Excess reagent, *184*
Exchange reaction, *162*
Exponential notation, 307–309

Fahrenheit temperature, 18
Fire, 36
Formula weight, 119, *138*

Gas, 36, 201–202, 204–205, *207*
Gay-Lussac's law, 288, *298*
General gas law, 281–287, *298*
Glucose, 135–136
Gram, *27*

Gram-atom, 125
Gram atomic weight, 124–125, *139*
Gram molecular weight, 127, *139*
Greek (ancient) concepts of matter, 36
Group, 72, *73*
   of elements, 43, 46, *47*

Helium, 65
Heterogeneous material, 6, *27*
Homogeneous material, 5, *27*
Hydrogen, 63, 64, 68
Hydrogen chloride, structure, 85, 86
Hydronium ion, 237, *250*
Hypothesis, 7, *27*

Inert gas, 37, *47*, 67
International prototype meter, 9
Ion, 90, 91, *95*
Ionic bond, 91
Ionic compound, 88, 89, *95*, 119
Ionic equation, 244, 245, *250*
Ionic solid, *208*
Isotopes, 63, 67–70, *73*

Kelvin temperature scale, *298*

Kinetic molecular theory, 202–204, 206, *208*

Law, 6, *27*
Law of conservation of matter, 55, *73*
Law of definite composition, 56, 57, *73*
Law of LeChatelier, 265, *271*
Limiting reagent, 181–183, *184*
Liquid, 36, 201, 205, *208*
Liter, *27*
Litharge, 108
Lithium, 69
Lithium chloride, 70
Litmus, 237, 239, *250*
Lord Rutherford, 60
Lye, 108

Magnesium chloride, structure, 89
Mass number, 69, *73*
Matter, 3, *28*
Melting point, *208*
Mendeleev, Dimitri, 45
Metallic element, 37, *47*
Metallic solid, 200, *208*
Meter, 9, *28*

Metric system, 8, *28*
   units of linear measurement, 11
   units of volume measurement, 12
   units of weight measurements, 13
Milk of magnesia, 108
Millimeter, 10
Mixture, 5, *28*
Molar volume, 296, *298*
Molarity, 218–219, *227*
Mole, 123–130, *139*
Mole ratio, 172, *185*
Molecular formula, 130, 133, 135, 138, *139*
Molecular ratio, 171, 172, *185*
Molecular solid, 199, 200, *208*
Molecular weight, 118, *139*
Molecule, 86, 90, *95*
Molecules, shapes of, 316, 317

Net ionic equation, 245, *250*
Neutralization reaction, 241–243, *250*
Neutron, 60, 63, *74*
Nomenclature, 102
   additional, 318, 319
   prefixes in, 103, *109*
   stock system, 106–107, *109*
   suffixes in, 102, *109*

Nonmetallic element, 37, *47*
Normality:
   of an acid, 246–247, *251*
   of a base, 247, *251*
Nuclear change, 5
Nucleus, 61, 62, *74*

Orbitals, shapes of, 316
Outer shells of electrons, 67
Oxides, nomenclature of, 318

Particulate nature of matter, 57
Percent by weight, *95*, 222–224
Percent composition, 120
Percentages by weight, 120
Period, 72, *74*
   of element, *47*
Periodic table, 43–45, *47*
pH, 269–270, *271*
Plücker, J., 59
Polyatomic ion, 91–92, *95*, 104–105
Polyprotic acid, 243, *251*
Powers of ten, 10
Pressure, 203–204, *208*
Properties, *28*
Proton, 60, 63, *74*

Quick lime, 108

Ratio-proportion method, 179–181
Reactants, *162*
Reaction products, 151, *162*
Reversible reaction, 264, *271*

Salt, 108, 240, 241, *251*
Saturated solution, 263
Scientific method, 6, *28*
Scientific notation, 10
Slacked lime, 108
Sodium, properties, 83
Sodium chloride, 46, 70
  properties of, 83, 84
  structure of, 88, 90, 200
Sodium hydroxide, structure, 91
Solid, 36, 198, *208*
Solute, 214, *227*
Solution, 214, *227*
Solvent, 214, *227*
Specific gravity, 25, *28*
Standard atmosphere, *298*
Standard temperature and pressure (STP), *298*

Stock system of nomenclature, 106–107, *109*
Strength of an acid, 266
Strong acid, 266, *271*
Strong base, 267, *271*
Strong electrolyte, 241
Structure:
  of ice, 199
  of sodium chloride, 200
Sublimation, 205, *208*
Substance, 4, *208*
Sulfate ion, 92
  structure of, 92
Surface tension, 201
Symbols of elements, 38, *48*
  derived from English names, 40, 41
  derived from foreign names, 39

Temperature, 18
Temperature conversion:
  from Celsius to Fahrenheit, 21
  from Fahrenheit to Celsius, 19, 20
Theory, 7, *28*
Thomson, J. J., 59

Unit analysis, 171
Universal gas constant, *298*

Valence, 87, *95*
Valence electrons, 86, *95*
Valence shell, 67, *74*
Vapor pressure, 206

Water, 108, 198, 199
   structure of, 86
Weak acid, 266, *271*
Weak base, 266–267, *271*
Weight, 3